Research Ethics and Social Movements

W0113779

What ethical challenges are faced by researchers studying social and political movements? Should scholars integrate their personal politics and identities into their research? What role should activists have in shaping the purposes or processes of social scientific research? How do changing political contexts affect the ethical integrity of a research project over time?

These are some of the live issues of research ethics that face students and scholars whose research 'subjects' are located in contentious political terrain. The contributors to this volume expose their own ethical thinking as they have met such challenges head on. Each explores real dilemmas of ethical practice on the ground as they carry out research on social movements across the globe. Authors examining pro-democracy activists in Malaysia, sanctions-breakers in the Gaza Freedom Flotilla, environmental health organisations in North America and much else find that the narrow confines of research ethics committees and institutional review boards offer little guidance on the questions that really matter. They offer instead a demonstration of continual reflexivity that is both personal and political in its approach. This book opens up debate on research ethics, delineating key challenges and offering hopeful and practical ways forward for real-world, ethical social science.

This book was originally published as a special issue of *Social Movement Studies*.

Kevin Gillan's publications have focused on the generation and communication of ideas in social and political movements and on their relationships with technology and globalisation. His latest research project, *Making Corporations Moral*, examines normative justifications for a range of activities that challenge corporate behaviour.

Jenny Pickerill is a Professor of Human Geography at Sheffield University, UK. She has worked with a diverse range of activists, from anti-roads protestors in Britain to Indigenous environmental activists in Australia and eco-villagers in Thailand.

Research Ethics and Social Movements

Scholarship, Activism and Knowledge Production

Edited by
Kevin Gillan and Jenny Pickerill

LONDON AND NEW YORK

First published 2015 by Routledge

2 Park Square, Milton Park, Abingdon, Oxon, OX14 4RN
605 Third Avenue, New York, NY 10017

Routledge is an imprint of the Taylor & Francis Group, an informa business

First issued in paperback 2020

British Library Cataloguing in Publication Data
A catalogue record for this book is available from the British Library

ISBN 13: 978-1-138-82076-0 (hbk)
ISBN 13: 978-0-367-73912-6 (pbk)

Typeset in Times New Roman
by RefineCatch Limited, Bungay, Suffolk

Publisher's Note
The publisher accepts responsibility for any inconsistencies that may have arisen during the conversion of this book from journal articles to book chapters, namely the possible inclusion of journal terminology.

Disclaimer
Every effort has been made to contact copyright holders for their permission to reprint material in this book. The publishers would be grateful to hear from any copyright holder who is not here acknowledged and will undertake to rectify any errors or omissions in future editions of this book.

Contents

CONTENTS

Citation Information

The chapters in this book were originally published in *Social Movement Studies*, volume 11, issue 2 (April 2012). When citing this material, please use the original page numbering for each article, as follows:

Chapter 1
The Difficult and Hopeful Ethics of Research on, and with, Social Movements
Kevin Gillan & Jenny Pickerill
Social Movement Studies, volume 11, issue 2 (April 2012) pp. 133–144

Chapter 2
Social Movements and the Ethics of Knowledge Production
Graeme Chesters
Social Movement Studies, volume 11, issue 2 (April 2012) pp. 145–160

Chapter 3
Reflexive Research Ethics for Environmental Health and Justice: Academics and Movement Building
Alissa Cordner, David Ciplet, Phil Brown & Rachel Morello-Frosch
Social Movement Studies, volume 11, issue 2 (April 2012) pp. 161–176

Chapter 4
Ethical and Political Challenges of Participatory Action Research in the Academy: Reflections on Social Movements and Knowledge Production in South Africa
Marcelle C. Dawson & Luke Sinwell
Social Movement Studies, volume 11, issue 2 (April 2012) pp. 177–192

Chapter 5
The Gaza Freedom Flotilla: Human Rights, Activism and Academic Neutrality
Anne de Jong
Social Movement Studies, volume 11, issue 2 (April 2012) pp. 193–210

Chapter 6
Sisterhood and After: Individualism, Ethics and an Oral History of the Women's Liberation Movement
Margaretta Jolly, Polly Russell & Rachel Cohen
Social Movement Studies, volume 11, issue 2 (April 2012) pp. 211–226

Chapter 7
Ethics, Activism and the Anti-Colonial: Social Movement Research as Resistance
Adam Gary Lewis
Social Movement Studies, volume 11, issue 2 (April 2012) pp. 227–240

Chapter 8
Disclosed and Willing: Towards A Queer Public Sociology
Ana Cristina Santos
Social Movement Studies, volume 11, issue 2 (April 2012) pp. 241–254

Chapter 9
Asking Tough Questions: The Ethics of Studying Activism in Democratically Restricted Environments
Sandra Smeltzer
Social Movement Studies, volume 11, issue 2 (April 2012) pp. 255–272

Chapter 10
Research Note: A Personal Reflection on Negotiating Fear, Compassion and Self-Care in Research
S.J. Creek
Social Movement Studies, volume 11, issue 2 (April 2012) pp. 273–277

Please direct any queries you may have about the citations to
clsuk.permissions@cengage.com

Notes on Contributors

Phil Brown is University Distinguished Professor of Sociology and Health Sciences, and director of the Social Science Environmental Health Research Institute at Northeastern University. He began working on environmental health in the mid 1980s when he wrote *No Safe Place: Toxic Waste, Leukemia, and Community Action*, about the Woburn childhood leukemia cluster. His other books are *Toxic Exposures: Contested Illnesses and the Environmental Health Movement, Perspectives in Medical Sociology, Illness and the Environment, Social Movements in Health*, and *Contested Illnesses: Citizens, Science and Health Social Movements*. His current research includes biomonitoring and household exposure to chemicals and particulate matter, social policy and regulation of flame retardant chemicals, and techniques and ethics of reporting data to study participants. Phil co-directs the Community Engagement Core and Research Translation Core of Northeastern University's Superfund Research Program. He directs an NSF training grant, 'New directions in environmental ethics: Emerging contaminants, emerging technologies, and beyond', that supports doctoral students and postdocs.

Graeme Chesters is Senior Research Fellow in the Department of Peace Studies and Humanities at the University of Bradford, UK. He is author (with Ian Welsh) of *Complexity and Social Movements* and *Social Movements: The Key Concepts* and was a member of the editorial collective who produced *We Are Everywhere: The Irresistible Rise of Global Anticapitalism*.

David Ciplet is a doctoral candidate in the Department of Sociology at Brown University. His research focuses on processes of social change in climate change politics. His recent articles include, 'Contesting Climate Injustice' and 'The Politics of International Climate Adaptation Funding', both in *Global Environmental Politics*. He is co-author of the book manuscript, *Power in a Warming World: the New Geopolitics of Climate Change*, contracted by MIT Press.

Rachel Cohen is the Research Fellow for 'Sisterhood and After: The Women's Liberation Oral History Project'.

Alissa Cordner is Assistant Professor of Sociology at Whitman College. She received her PhD in sociology from Brown University. Her research focuses on environmental sociology, the sociology of risk, environmental health and ethics, civic engagement, and public engagement in science and policy making. She is currently completing a book manuscript about the intersection of science, regulation, activism, and industry decision making related to environmental health risks and consumer exposure to chemicals.

S.J. Creek is a visiting Assistant Professor of Sociology at Hollins University, USA, whose research interests lie at the intersection of religious movements, identity and sexuality.

Marcelle C. Dawson works as a Senior Researcher attached to the South African Research Chair in Social Change at the University of Johannesburg, South Africa. She obtained a DPhil in Politics from the University of Oxford, UK, in 2008. She runs a research project on social movements and democracy and is currently working on a monograph based on her thesis on the Anti-Privatisation Forum and struggles against the commodification of water in South Africa. Dawson is a board member of the International Sociological Association's Research Committee 47 (Social Classes and Social Movements) and convenes a working group on social movements and popular protest for the South African Sociological Association.

Kevin Gillan's publications have focused on the generation and communication of ideas in social and political movements and on their relationships with technology and globalization. His latest research project, Making Corporations Moral, examines normative justifications for a range of activities that challenge corporate behaviour.

Margaretta Jolly is author of *In Love and Struggle: Letters in Contemporary Feminism* (2008) (winner of Feminist and Women's Studies Association Book Prize), and editor of *The Encyclopedia of Life Writing* (2001) and *Dear Laughing Motorbyke: Letters from Women Welders of the Second World War* (1997). She is director of the University of Sussex's Centre for Life History and Life Writing Research, UK.

Anne de Jong (PhD) is an anthropologist specialized in violence, human rights and nonviolent social movements in the Middle East. Focusing on Israel and the Occupied Palestinian Territories, she strongly rejects the assumed neutrality of academia and activism as separate spheres of conduct. As such, her work aims to bridge the divide between research-based knowledge, policy making and social change from below. Authoring two books in 2011, de Jong currently works on a follow-up study deepening the conceptualizations of social justice and human rights among those opposing violent oppression.

Adam Gary Lewis is a settler anarchist working towards settler decolonization in social movement and academic contexts. He is currently working on a PhD at York University in Environmental Studies on anarchist engagements with Indigenous struggles of resistance and the possibilities of developing anarcha-Indigenism as a form of anti-authoritarian and anti-colonial politics. Adam works towards decolonization with GRIS (Grand River Indigenous Solidarity) in Kitchener where he lives and serves as a co-editor of *Affinities: A Journal of Radical Theory, Culture, and Action*.

Rachel Morello-Frosch is a professor in the Department of Environmental Science, Policy and Management and the School of Public Health at the University of California, Berkeley. Her research integrates environmental health science with social epidemiologic methods to assess potential synergistic effects of social and environmental factors in environmental health disparities. In collaboration with scientific colleagues, she has developed scientifically valid and transparent tools for assessing the cumulative impacts of chemical and non-chemical stressors to inform regulatory decision-making to advance environmental justice goals. She is assessing the application of these methods for implementation of climate change policies in California. She has also written about the scientific challenges and bioethical considerations associated with exposure assessment and chemical biomonitoring research in economically and racially marginalized communities. She is co-author of *Contested Illnesses: Citizens, Science and Health Social Movements* (with Phil Brown and Steve Zavestovski). Her research is supported by the National Institute of Environmental Health Sciences, National Science Foundation, U.S. EPA, Cal-EPA and numerous private foundations.

Jenny Pickerill is a Professor of Human Geography at Sheffield University, UK. She has worked with a diverse range of activists, from anti-roads protestors in Britain to Indigenous environmental activists in Australia and eco-villagers in Thailand.

Polly Russell is Lead Content Specialist in Social Science Collections and Research at The British Library, UK.

Ana Cristina Santos is a Sociologist. She holds a PhD in Gender Studies, University of Leeds, UK and an MA in Sociology, University of Coimbra, Portugal. She is a Senior Researcher at the Centre for Social Studies, University of Coimbra, and Honorary Research Fellow at the Birkbeck Institute for Social Research, University of London, UK. Significant publications have been published in Brazil, Portugal, the UK and the USA. She is currently writing a monograph entitled *Social Movements and Sexual Citizenship in Southern Europe*. She is also a feminist and LGBT and queer activist.

Luke Sinwell completed his MA (2005) and PhD (2009) at the University of the Witwatersrand in Johannesburg, South Africa. He is currently a Post-Doctoral Fellow with the South African Research Chair in Social Change at the University of Johannesburg, South Africa. His research interests include the politics and conceptualisation of participatory development and governance, social movements and housing struggles, non-violent and violent direct action as a method to transform power relations, ethnographic research methods and action research. He is currently involved in a project called 'The Voices of the Poor in Urban Governance: Participation, Mobilization and Politics in South African Cities'.

Sandra Smeltzer's primary areas of research and publication include communication in transitioning and developing countries, the scholar—activist dialectic, ICTs for social justice and alternative media pedagogy. She is currently working on a Social Sciences and Humanities Research Council of Canada-funded project examining the use of Web 2.0 technologies for resistance in Southeast Asia (with a particular focus on Malaysia). In 2011, Dr Smeltzer was awarded Western University's inaugural Humanitarian Award for her international work and was named one of Canada's Top 25 Most Influential Women by the Women of Influence magazine.

The Difficult and Hopeful Ethics of Research on, and with, Social Movements

KEVIN GILLAN* & JENNY PICKERILL**
*School of Social Science, University of Manchester, Manchester, UK, **Department of Geography, University of Leicester, Leicester, UK

ABSTRACT   *This article explores a number of key questions that serve to introduce this special issue on the ethics of research on activism. We first set out the limitations of the bureaucratic response to ethical complexities in our field. We then examine two approaches often used to justify research that demands time consuming and potentially risky participation in research by activists. We label these approaches the ethic of immediate reciprocity and the ethic of general reciprocity and question their impacts. We note, in particular, the tendency of ethics of reciprocity to preclude research on 'ugly movements' whose politics offends the left and liberal leanings predominant among movement researchers. The two ethics also imply different positionalities for the researcher vis-à-vis their subject movement which we explore, alongside dilemmas thrown up by multiple approaches to knowledge production and by complex issues of researcher and activist identities. The overall move to increasing complexity offered by this paper will, we hope, provide food for thought for others who confront real-world ethical dilemmas in fields marked by contention. We also hope that it will encourage readers to turn next to the wide range of contributions offered in this issue.*

Introduction

Every stage of the research process into social movements can introduce complex ethical questions. The issues we choose to address are often highly politicised and involve our own moral judgements and sympathies. The groups and individuals with whom we engage, whether directly or through documentary records, may be in positions of peculiar vulnerability. They may be relatively powerless by virtue of their social situation, their activities may be covert or illegal, and they may face a high risk of repression. The data we gather, then, have special risks associated with them (Blee and Vinning, 2010), but ethical challenges do not stop once we insert our own analyses. Rather, we must make choices about what we report, in what terms we report it and what we leave unsaid, judging the risks faced by research respondents and deciding on the importance of giving voice to those who feel under-represented in their societies. Moreover, we must choose which audiences we wish to address. These issues and many more are likely to be familiar to anyone who has engaged in research on social movements, whatever the particular methodological techniques they employ. While some of these ethical challenges may seem unique to the study of social movements, we also believe that the lessons available here may be much more broadly applicable to original research in a number of cognate fields.

This special issue was prompted in part by the rising demands for researchers to specify the ethical implications of their work; demands of which anyone working within a higher education institution in the USA, Europe and perhaps further afield, will be aware. In the first section below, we argue that the bureaucratic nature of ethical review processes offers a partial, and at times, debilitating approach to carrying out ethical research. A part of the problem is the fact that deferral to research ethics 'specialists' empties such processes of the important complexities resulting from the substantive characteristics of any research project. This introduction highlights a number of those complexities as commonly experienced in research relating to social movements. Such complexities are not necessarily best served by more complex and sensitive (and therefore onerous) review procedures; they will not dissolve in a bureaucratic solution. It is instead the practices of sharing experiences, airing conundrums and puzzling over problems in the community of scholars and research participants that will allow researchers to continually improve the ethical standards of data collection, analysis and dissemination. This special issue is intended to be one step in that direction. No field has sharply defined boundaries of course, and it remains important to consider problems and solutions found in a range of social scientific fields; we hope, in this regard, that this special issue will have a relevance for anyone whose research subject is marked by contention or conflict, or who recognises the urge to take a principled stand for, or against, those who have been involved in or benefit from their research.

Here, we explore a few themes of particular interest that cut across the papers in this special issue, and a couple which perhaps have yet to be adequately addressed. In the next few paragraphs, we will briefly outline the role of institutional ethical review before this introduction (and indeed the whole volume) opens out the notion of research ethics to much broader and important questions. We conclude by identifying some hopeful ways forward in practising an ethical approach to researching activism.

Bureaucratic Demands and 'Real' Research Ethics

Not so long ago, demonstrating awareness of the relevant disciplinary 'code of conduct' for research was often all that was required to reassure supervisors, funders or managers that the ethical implications of one's social science research project had been thoroughly thought through. Research ethics committees in many institutions now have considerably strengthened oversight, lengthy forms and demanding panel meetings in which one must defend the procedures used for research. Additional demands are thus placed not only on professional researchers but also on our doctoral students and, in some institutions, undergraduates too. There are undoubtedly instances where review processes will have encouraged better practices and one positive outcome is the increasing focus on research ethics in our teaching. It has now become rather more automatic to consider the genuine ethical quandaries that might arise in one's research project although, in our own experience, and that of many of our students, the 'real ethics' that we considered before submitting our projects to ethical review never seem to have a box on the form.

While drafting this introduction we received an email publicising a book that promised to help us 'Avoid Ethical Lapses that Put Your Project and Organization at Risk' with 'the use of graphics, checklists, examples, and other tools' (marketing materials for Kliem, 2011). The check-box approach to ethics, which is the mainstay of any bureaucratic procedure, may help deal with certain sorts of risk but is ultimately limited. By reducing ethics to a generic checklist, often not even sensitive to differences between physical and social sciences, let alone different disciplines, much of the ambiguity of ethics is lost. Yet it is precisely the ambiguity—the uncertainty of what research is for, who it should benefit and what risk and harm is acceptable in conducting research—that ethical thinking should help us navigate. Ethical questions are not clear cut but are heavily contextualised by the researchers' own positionality and relationship to research subjects. Instead many university ethical procedures simply require social scientists to ensure written consent forms and project information sheets are used. Consent forms are required for ethical approval, regardless of the fact that they can be highly inappropriate for some forms of activism research. This process of approval is further generalised (and thus limited) by the process of peer approval operating within specialised departments which means that those who determine if research is ethical often have no knowledge of the field of research which they judge. We have encountered ethics officers who resist any alternative approaches because they have no knowledge of social movements or the variety of approaches taken in researching with them.

Reducing all of this complexity to questions on whether a 'gatekeeper' will need to provide access or whether physical tests will be carried out on subjects ignores the importance of the *process* of research, the need to navigate scenarios where there is no clear 'right' choice, and the requirement to respond to changing conditions. At many universities in Britain, although ethics committees have been established, there is still no mechanism for checking whether researchers actually implemented the ethical approach they signed up to. There is no feedback mechanism as to how research was actually undertaken or any reflection or sharing of lessons learnt in the field. In this way, ethics are still perceived as a static consideration to be completed early in any research project, rather than the dynamic, complex and ongoing dilemma that researchers really face. Indeed, in this volume many of the most difficult and complex ethical questions and choices faced by researchers emerged late in the research process (see both De Jong and Creek, this issue) when much data collection had already been completed. It is often the temporal implications of participation, the evolving affiliation with research subjects and the heightened politicisation of researchers (often through the research itself) that emerge which raise really interesting ethical dilemmas. As such these questions of participation, politics, identity, reciprocity and social justice are the issues too often missed from a bureaucratic approach to ethics.

The Complexity of Reciprocation

There is a growing trend within social movement research for academics to consciously, indeed loudly, take on the role of 'activist-scholars' (Routledge, 1996; Maxey, 1999; Fuller & Kitchen, 2004; Bevington & Dixon, 2005; Graeber, 2009; The Autonomous Geographies Collective, 2010). Such arguments tend to assert the potential for academics to make a real and positive impact on movements they are studying. This may be simply by using ethnographic methods that enable participation with movements that are being studied, thereby adding to the movement's number and offering a particular set of skills. Or it may be that the activist-scholars focus more on their particular skills, offering research and writing activity that may be beneficial to movement participants (though see Flacks, 2005), while not necessarily putting themselves on the line in physical actions. Taking on the activist-scholar role is often justified through an argument, we will refer to as the ethics of immediate reciprocation: participation in research has its costs for activists through the time taken and in the personal security risk (which may be perceived as considerable when the researcher is not personally known to participants beforehand), so, academics must ask, what can we do to return favours due? Directly aiding the movement one studies becomes one of the answers.

Such arguments are explored and advanced in a number of contributions to this special issue, so we will not detail them further here (see Dawson and Sinwell; Cordner *et al*.; Smeltzer; Chesters; and Santos; this issue). We would, however, like to sound a few notes of caution concerning this ethical claim. We want to be cautious about advocating reciprocity as a simple and easy resolution to the inequity of power between researchers and research participants, as a way to gain access to groups, as a way to enact social change, or to share in the 'risk' of activism *per se*. While both of us have practised reciprocity in various ways over the years, it has rarely been an easy ethical approach.

The main issue with immediate reciprocity is that problems of objectification do not disappear through participation (Roseneil, 1993). As the trend for identifying as activist-scholars has grown so too has the kudos for being an activist within academia. No longer marginalised for being subjective, this active engagement with social change actors has meant that it is increasingly possible to benefit from links with social movements in order to further one's academic career (most notably in Britain through the inclusion of an 'impact' factor in our formal Research Excellence Framework which measures our engagement with those 'outside' academia). Benefitting from others' knowledge and actions to further one's career is the antithesis of what an ethical approach should be about (see also Chesters, this volume). At the same time, there may be limits to how useful the activity offered by academics really is; as Croteau has it, 'Becoming an academic to support social movements is akin to launching a space program to develop a pen that writes upside down. At best, it is a circuitous route' (2005: 20). We do not make these points as an accusation against those scholars we reference above and recognise that there are other justifications for the activist-scholar role. It is noteworthy, however, that even a perception of an instrumental, exchange-based approach might have longer-term dangers for relationships between social scientists and activists.

Another result of an emphasis on reciprocation is to narrow those movements which we study. Most of the arguments concerning immediate reciprocity have come from scholars studying movements that are, broadly speaking, of the left. Given a decided leftward lean among the social movement researchers, activist-scholars can expect some argument concerning the particular politics of particular movements, but are unlikely to face hostility about the broader ambitions of the movements they attempt to assist. However, it is intellectually essential not to restrict our collective endeavours to research on movements with which we can have such easy relationships. It is necessary to research anti-abortion movements, racist national movements, terrorist movements and the whole gamut of fundamentalist religions. How, then, does one apply the ethic of immediate reciprocation when one requests (as Creek has done, this issue) the participation of ex-gay activists who proselytise on the basis that people need redemption from their non-heterosexual identities? Genuine attempts to assist ugly movements achieve their goals would likely meet consternation from those who have argued so strongly for 'embedded' activist-scholarship.

If the study of a diversity of movements is important for social movement scholarship, then the ethic of immediate reciprocation raises uncomfortable demands and introduces political contestation squarely at the heart of the collective, intellectual endeavour. While it would be misleading to pretend that individual academics' politics could be banished from their research, to turn all research on activism into political action would be to weaken any claims to the systematic creation and critique of knowledge that the academic field could muster.

Thus, there is a need to critically examine whether immediate reciprocation could preclude the important task of exploring a diversity of social movements, and whether the ethic of immediate reciprocation could itself become a dominant and ultimately unhelpful dogma in social movement studies. There is a danger that reciprocation is practised simply as a way to gain access, rather than as a genuine desire to aid the movement; potentially an ethically dishonest approach. The growing assumption that reciprocation is a preferred ethic has also had unfortunate implications on activists' expectations of researchers. Reciprocation becomes particularly messy and emotionally fraught when research subjects assume abundant resources are available, or request help with tasks in which the researcher has no skill (The Autonomous Geographies Collective, 2010; see also Dawson and Sinwell, this issue). Though as Smeltzer (this issue) argues often the most useful immediate reciprocation involves 'back office' work which is less visible but just as important as front line direct action. These issues are in addition to the long-known quandaries of when such participation and reciprocation blurs the ability of academics to think critically about their subject and the need for continual reflexivity about our roles and positionality (England, 1994; Plows, 1998; Cordner et al., this issue).

The Value of Academic Knowledge Production and Dissemination

The desire to reciprocate on research participants' investment in a project is hardly the only justification for activist-scholarship. To the extent that individuals mobilise the knowledge and skills of their intellectual discipline in the service of particular groups, then the activities of activist-scholars may fit with calls for a more public sociology (which may, of course, be extended to many social science disciplines) and, perhaps more surprisingly, with the desire of funding bodies to see the impact of research on a variety of stakeholders or 'user groups' relevant to the study.[1] These arguments imply an older ethic of general reciprocity: it may be impossible to 'pay back' a research participant for their efforts in any direct way that does not somehow compromise the research, but the utility of the results of research will offer some benefit at a broader, societal level.

The main dilemma then becomes whether, and how, knowledge produced should be of relevance to social movements. Knowledge produced should at least be physically accessible to those who participated in its creation, though so much, including unfortunately the papers in this journal, remain locked up in costly or inaccessible publications (Pickerill, 2008). Even if such publications are freely shared, however, the language, findings and timeliness of our research can be of limited use to social movements.

Activists are often very interested in research that addresses the particular problem they are mobilising around, which may sit outside the domain of social movement studies, or in studies that offer lessons about strategy and tactics. Within this journal's main field, it may be work on movement outcomes and impacts that most obviously pique the interest of activists along, perhaps, with practical insights into the processes of mobilisation. But many areas of scholarship may have huge merit academically speaking, but remain of little direct interest to activists. Indeed, if social movement research projects rely mostly on listening to activists analyses and then simply parroting these lessons to an academic community, then it would be difficult to see it as having any benefits to the social movements themselves beyond, perhaps, amplifying the voices of activists.

Academia, however, also offers particular value in knowledge production as a consequence of its claims to rigour and systematicity. Academic research includes (or at least ought to include) a very significant analytical step such that the sum of the parts is greater than simply a collection of data drawn from the goodwill of movement participants as research subjects. Moreover, social movement scholars can, and often do, work in other fields, making significant contributions to knowledge about neoliberal globalisation, patriarchy, racism and a whole host of movement-relevant issues. As Don Mitchell has persuasively argued, 'to make a difference beyond the academy it is necessary to do good and important, and committed work *within the academy*' (2004, p. 23). Thus, the ethical question remains as to whether it is enough, having elicited 'rich data' from our research respondents, to use these solely in pursuit of intellectual plaudits in the academy. This argument from general reciprocity, resting on the assumption that academic knowledge has a particular role alongside and interacting with activist knowledge, may be an ethical justification for inviting activists to behave as research subjects. In doing so, it seeks to avoid risking the appearance of condescension to movement participants (by telling them what they already know, just wrapped up in academic language) and excessive political involvement to our scholarly peers.

Yet, as discussed by both Santos and Chesters (this issue) many of us have come to accept that all our knowledge is situated knowledge, and that whether we are open and honest about it or not we write with a subjectivity that is often formed by an embedded and sympathetic position to social movements. In recent years, the validation of the participatory approach to social movement studies has supported the claim that knowledge that is grounded and situated within movement praxis is more worthwhile, perhaps more authentic, than knowledge developed in the academy. This scholarship can actively situate itself against capitalism, corporations and elites and in support of those social movements with which we work (Barker, 2008). While broadly supportive of the value of this situated approach, there is a danger of losing sight of the different values of different kinds of knowledge. Building a hierarchy with 'authentic' or 'situated' knowledge at the top may undermine the value of knowledge developed in the academy for academic ends. Indeed, there remains a responsibility on academics to clearly articulate and advocate the value of knowledge-in-general and thus the wider purposes of academic work; that is, the value of a general reciprocity approach.

These arguments are not intended to undermine the importance of embedded participatory research—especially those projects where the goals of the research are co-designed by academics and participants (see Cordner *et al.*, this issue)—or to undervalue those who seek to produce and disseminate socially useful knowledge, especially in easily understandable forms. However, we do oppose any dogma that one form of knowledge production is superior to all others. Ethical questions as to who the research should be useful to, and for what ends, do not require disregarding the importance of producing knowledge-in-general (Chase, 2003). Rather, these questions should make us alert to the risks of how knowledge might be misused (such as for state surveillance), and to navigate the necessary attempts at co-production, but also to create space and respect for knowledge production in other forms.

Identity, Dualisms and the Spatial Politics of Responsibility

One of the main themes emerging from the papers in this volume is the acknowledgement that our identities (as researchers, activists, women, white, homosexuals, and so on) overlap and intertwine with our research and, in practice, negate many of the dualisms which have been long established in academia. Moreover, these identities are dynamic and fluid and thus often co-evolve with our research. Ethically, this complicates our positionality but also strengthens our understanding of our subjectivity and involvement in knowledge production. At the same time, for many who conduct research in places far from our homes and academic institutions, this assertion of our identity and visible insertion of ourselves into our research can help us better understand how our sense of responsibility to our research subjects stretches and changes over time and space (Smeltzer, this issue). In other words, for many of us there is a spatial politics of responsibility to those social movements with whom we work, but this responsibility can be difficult to articulate and enact within the traditional framework of academia (Massey, 2004).

This acknowledgement of the importance of our deeply embedded understanding of our identity moves the ethical debates beyond simply a question of what form of reciprocation is appropriate. Rather it raises fundamental questions about our sense of self, when our work might at times clash with the increasingly capitalist and corporate university institutions in which many of us work (Barker, 2008; Dawson and Sinwell, this issue). Ethical considerations then become about much more than bureaucratic checklists of practical elements we must include in our research, they become (and always were) about how we understand ourselves, our role in social change and our very identities.

Celebrating and being open about our identities and their complexity then becomes a core ethical approach to research activism. The 'rewarding intersection' between ourselves as social actors and academics should be celebrated (Santos, this issue). An emphasis on uniqueness and diversity, rather than a single route to ethical research, echoes the way we would expect activists to favour a variety of strategies and tactics for social change. This is an extension of the long-standing work on the need for reflexivity (England, 1994), but it is more overt in its discussion of who the researcher is and how that interplays with our work. To ignore the importance of identity to ethics risks failing to value passion about the movements with whom we work as much as we value rigour in our methods and findings.

A focus on identity also helps overcome a perception that ethical dilemmas need only be examined by early career researchers. Too often ethics are deemed something doctoral students need to overcome and much reflection and writing on ethical dilemmas is by early career academics. It appears that once an ethical approach has been determined then academics can get on with the 'real' research and ethical questions are rarely revisited. If we are to take identity and ethics seriously, then ethical questions become ongoing and central to our research. This better reflects the ongoing physical and emotional labour involved in social movement research. Ethics need to include a reflection on care-of-self as well as care of research participants, especially when one is physically in the front line (De Jong, this issue) or emotionally exposed (Creek, this issue). This individualisation of ethical problems prevents us from adequately sharing success stories and showing solidarity to those who have taken difficult ethical paths. It is too often left to the individual to justify why, as De Jong did, they ultimately participate, at great personal risk, in the social movements they were studying. We hope that this volume will go some way to sharing the stories of ethical issues and approaches and move towards a more collective sense of responsibility for these ethical dilemmas.

Overcoming the dualisms inherent in much research (such as academic-activist, writer-social change actor, elite-subject, university-society) also enables us to articulate the spatial politics of responsibility. Taking an ethical approach requires us to understand how, even when those with whom we work might be many thousands of miles away, we respect and incorporate and feedback our work. It requires us to understand how distance—physical and intellectual—does not negate our responsibilities.

A spatial politics also enables us to understand and situate social movements within broader conceptual frames—such as capitalism, colonialism (Lewis, this issue) or patriarchy—which in turn identifies hierarchies of power, the position of academics within those hierarchies, and may also help us avoid repeating past mistakes in the objectification and disempowerment of research subjects. By situating the academy as a space within these conceptual frames, it can be understood as an important space of social change itself. This helps us recognise the potential isolation of the academy's 'ivory tower', standing intellectually dislocated from wider society, while at the same time the institutional demands on the university embed it within systems of capitalism and colonialism. As Dawson and Sinwell argue in this issue, these processes reinforce the need to link our work on social movements with making changes within our own institutions.

Hopeful Ways Forward

Thus far, we have tended towards identifying the problems with existing ethical approaches to the study of activism and the complexity of attempts to navigate the ethical minefield. We would like to end by identifying some of the more successful strategies and hopeful ways forward—many of them illustrated in this volume.

A key part of an ethical research process is recognition of our location within the broader dynamics of society and seeking to reflexively critique and adjust that positionality. Lewis (this issue) does this by asserting a role for academics, particularly those working with Indigenous activists resisting colonialism. This involves acknowledging the need to recognise oppression and domination in places where it might have previously been invisible and seeking to address it. Similar approaches have been made in overcoming patriarchy and gender domination and in the realm of sexuality (Santos, this issue). Smeltzer (this issue) also demonstrates a sensitive approach to recognising the impact of power differentials, in her case with a focus on the limited freedoms of activists (both as activists and as research participants) in democratically restricted societies. Both authors, in their different ways, indicate the importance of research that moves further afield than the norm, seeking to understand the importance of different political and social contexts. We would certainly want to encourage other attempts to find broader perspectives and note that studies of movements based in African and Asian countries are particularly lacking in the field (though see Smeltzer and Dawson and Sinwell, this issue).

While we have presented some cautionary notes about the ways in which different forms of knowledge can interact, it is clear that a number of scholars are working hard to include

8

movement actors within the knowledge production process. Cordner *et al.* (this issue) reflect on some of the dilemmas introduced here and demonstrate the relevance of community-based participatory research methods to social movements. Chesters (this issue) advocates, and has long practised, co-production of knowledge and, on a practical level, collaborative working with social movements and co-writing. This approach is time consuming and it can be difficult to negotiate the different agendas involved, but ultimately it can produce ethically robust and important knowledge. Moreover, as de Jong argues, taking adequate stock of the knowledge and perspectives of the movements we study may be essential in confronting some long-held biases on the academic mainstream. While this work often assertively amplifies the activist voice within all writings, others such as Jolly, Russell and Cohen (this issue) argue for the importance of telling the stories of social movements through individual voices: making the personal political, situating knowledge within personal trajectories and journeys and asserting the importance of oral history. Too often the voice and identity of individual activists is subsumed into a broader narrative and the nuances and complexities of their journey are lost. Jolly and colleagues offer an approach through oral histories and explore its ethical opportunities and limitations.

In addition to the approaches reflected on in the rest of this volume is the growing emphasis on taking our work outside the academy. Works like that of Mitchell (2008) seek to celebrate attempts to engage beyond academia and co-produce useful political work. Authors such as Fuller and Kitchen (2004) have long argued for a subversion of academic structuring of knowledge through peer-reviewed publishing systems and government research assessment schemes. Increasingly, there are new forms of solidarity for academics who wish to enact alternative ethics in their research with social movements and to publish in different and varied forms, often with open access. Perhaps, however, there remains a need to shift ethical questions from the personal and individual to more collective and mutually supportive ways of understanding our ethical obligations and responsibilities. We hope that this collective sense of responsibility for ethics might begin to emerge from this volume, though really this is a continuation of a long-running debate and we intend for that debate to flourish further.

We have deliberately chosen a diversity of ethical dilemmas and social movements to populate this volume and these offer a range of perspectives. We hope this issue will become both an essential point of reference for researchers in our field and also a valuable set of reflections for all academics occupied with research in sensitive or complex social environments. As an academic field we are increasingly aware of the ethical dilemmas involved in our research. However, this awareness has also shaped our research topic choices and it should not put us off from tackling any complex social movement research. Perhaps it needs saying that as a journal we welcome research on the ugly movements described above, on movements of the right and on movements in majority world countries—we see these lacunae as a weakness in our current publications. That said, a key undercurrent of this theme is an assertion that the ends can no longer justify any unethical means in our research. There are no excuses; instead there are plenty of productive, political and hopeful ways forward in research on, and with, social movements.

Note

1. On public sociology, see Burawoy (2004); also Santos, this issue. In relation to funding bodies there has, in the UK, been much debate over the so-called 'impact agenda', with varying understandings of 'impact'

developing in different spaces. However, the Economic and Social Research Council (ESRC) may be taken as one important benchmark, with its insistence on funding proposals including an outline of potential research users and the ways in which research may benefit them. It is a little difficult to imagine the ESRC taking a favourable view of proposals that include significant time for the researcher to take part in activism for the sake of reciprocating activists' involvement in the research. (There are, of course, real methodological justifications for extended participation.) However, where a relatively formally organised group exists that might have a beneficial role in the research project, either in an advisory capacity or as a recipient of dissemination efforts, this would seemingly be looked on positively.

References

Barker, M. (2008) Progressive social change in the 'Ivory Tower'? A critical reflection on the evolution of activist orientated research in US universities, Paper presented at Australian Political Studies Association Conference, 6–9 July, Brisbane, Australia.

Bevington, D. & Dixon, C. (2005) Movement-relevant theory: Rethinking social movement scholarship and activism, *Social Movement Studies*, 4(3), pp. 185–208.

Blee, K. M. & Vining, T. (2010) Risks and ethics of social movement research in a changing political climate, in: *Research in Social Movements, Conflicts and Change*, Vol. 30, pp. 43–71 (Bingley: Emerald Group Publishing).

Burawoy, M. (2004) Public sociologies: Contradictions, dilemmas, and possibilities, *Address to North Carolina Sociological Association*, Social Forces, June, pp. 1–16.

Chase, S. (2003) Professional ethics in the age of globalization: How can academics contribute to sustainability and democracy now? *Ethics, Place and Environment*, 6(1), pp. 52–56.

Croteau, D. (2005) Which side are you on? The tension between movement scholarship and activism, in: C. David, W. Hoynes & C. Ryan (Eds) *Rhyming Hope and History: Activists, Academics, and Social Movement Scholarship*, pp. 20–40 (Minneapolis, MN: University of Minnesota Press).

England, K. (1994) Getting personal: Reflexivity, positionality, and feminist research, *Professional Geographer*, 46, pp. 80–89.

Flacks, R. (2005) The question of relevance in social movement studies, in: C. David, W. Hoynes & C. Ryan (Eds) *Rhyming Hope and History: Activists, Academics, and Social Movement Scholarship*, pp. 3–19 (Minneapolis, MN: University of Minnesota Press).

Fuller, D. & Kitchen, R. (2004) Radical theory/critical praxis: Academic geography beyond the academy? in: D. Fuller & R. Kitchen (Eds) *Critical Theory/Radical Praxis: Making a Difference Beyond the Academy?*, pp. 1–20 (Vernon and Victoria, BC, Canada: Praxis (e)Press).

Graeber, D. (2009) *Direct Action: An Ethnography* (Edinburgh: AK Press).

Kliem, R. L. (2011) *Ethics and Project Management* (Boca Raton, FL: CRC Press).

Massey, D. (2004) Geographies of responsibility, *Geografiska Annaler*, 86 B, pp. 5–18.

Maxey, I. (1999) Beyond boundaries? Activism, academia, reflexivity and research, *Area*, 31(3), pp. 199–208.

Mitchell, K. (2008) *Practicing Public Scholarship: Experiences and Possibilities Beyond the Academy* (Oxford: Blackwell Publications).

Pickerill, J. (2008) Open access publishing: Hypocrisy and confusion in geography, *Antipode*, 40(5), pp. 719–723.

Plows, A. (1998) 'In with the in crowd': Examining the methodological implications of practising partisan, reflexive, 'insider' research, Unpublished MA thesis, University of Wales, Bangor.

Roseneil, S. (1993) Greenham revisited: researching myself and my sisters, in: D. Hobbs & T. May (Eds) *Interpreting the Field* (Oxford: Clarendon Press).

Routledge, P. (1996) The third space as critical engagement, *Antipode*, 28(4), pp. 397–419.

The Autonomous Geographies Collective (2010) Beyond Scholar activism: Making Strategic Interventions Inside and Outside the Neoliberal University, *ACME*, 9, 2, http://www.acme-journal.org/Volume9-2.htm

Social Movements and the Ethics of Knowledge Production

GRAEME CHESTERS

Department of Peace Studies, University of Bradford, Bradford, UK

ABSTRACT   *This article argues that whilst some academic disciplines have been richly innovative in both their approaches to, and work with movements, for the most part social movements are still considered as objects of knowledge for researchers and academics, rather than as knowledge producers in their own right. It considers the ethical and methodological dilemmas that arise from this situation and instead proposes an ethics of engagement with critical subjectivities whose contextual and situated knowledge is both independent of the academy and valuable in its own right.*

Introduction

In this article, I argue that the study of social movements within the academy retains an implicit positivism that is underpinned by the idea that we live in a 'social movement society' (Meyer & Tarrow, 1998). This idea is very appealing, suggesting as it does, that advanced industrial democracies have institutionalised social movement forms of representing claims to such an extent that they have become a conventional part of the accepted mechanisms for democratic participation. However, the legacy of this normative framework is that movements are largely perceived as objects of knowledge for academics, rather than as knowledge-producers in their own right. So whilst the knowledge they formulate about their areas of concern is evidently interesting for policy-makers, their ontological or epistemological frameworks are less so because they are already assimilated within the normative ontology of liberal democracy.

I will argue that this epistemological prioritisation of academic knowledge production *about* movements continues, despite the turn towards a phenomenological epistemology by theorists of social movements, who recognise movements as knowledge-producing subjects with whom the academy might engage in processes of co-production (Melucci, 1989, 1996b; Santos, 2003; Escobar, 2009). This is ethically problematic, not least because social movements are liable to be reduced to commodifiable objects of knowledge to enhance either an academic's career or reputation and/or their university's competitive standing (Federici, 2009).

Instead, I argue that social activism produces critical subjectivities whose contextual and situated knowledge is both independent of the academy and valuable in its own right (Conway, 2004; Casas-Cortés *et al.*, 2008; Cox & Fominaya, 2009). This requires those of us situated in the academy to examine the ontological and epistemological basis from which we engage with social movements and the methodological commitments such an approach might oblige. Such an approach can be suggestive of an ethics of engagement that emphasises relationality, reciprocity and an openness to causal mechanisms of 'becoming' that are outside liberal democratic strictures.

Consequently, the first half of this article seeks to situate the work of the influential social movement theorist Alberto Melucci (1996a) within a realist ontological tradition (Giddens, 1984; Bhaskar, 1989) that can help us to understand the knowledge-practices of social movements. It then moves on to consider a methodological approach to research that is sensitive to the differing knowledge-practices of academics, activists, activist-academics and activist researchers and concludes by offering some examples of the methodological innovations such an account might require to be ethically consistent, drawing upon anthropological and feminist critiques of classical ethnography.

This article also contends that the ontological and epistemological frameworks articulated and practiced by some contemporary social movements offer sophisticated examples of how to think about the production of knowledge and its relationship to practice, and highlights instances of these within the contemporary movement milieu. To begin with though, it is perhaps worth reminding ourselves why the knowledge-practices of social movements are crucial to our understanding of new ways of being and acting together, that might address the failure of the 'business as usual' of the neoliberal model (Harvey, 2010; Mattick, 2011).

The Contemporary Context and a Politics of Possibilities[1]

At the end of the first decade of the 21st century, we are faced by a number of interrelated and dynamically connected complex crises, all of which involve potentially dramatic tipping points that will impact upon human/ecological systems. These range from irreversible climate change, peak oil and crises of food and fuel production to market crashes and financial contagion.

Civil society and social movements are frequently at the forefront of knowledge generation about potential crises in human/ecological systems and can be conceived as critical sensors of systems moving from the edge of chaos towards more profound societal and environmental change (Johnson, 2002). In some instances, social movements are also capable of acting as facilitators of feedback systems that can avert or promote these processes through their generation of different lifestyles, resistance or adaptation strategies, and alternative social and economic policies (Urry, 2003; Chesters & Welsh, 2006). Despite this, formal political systems are frequently slow to respond to the production of new knowledge by social movements and the relationship between movements and political elites is often antagonistic due to the inherent difficulty of accommodating social movement activism within representative democracies, without the loss of institutional political capital.

Therefore, the motivation behind this paper is the observation that if we are going to respond in a just and sustainable way to the challenges of the complex and converging crises outlined above, we need to move beyond an analysis of movements that sees them solely as producers of knowledge about those things they campaign for, or against. Instead, we must recognise their capacity to develop alternative political imaginaries—a politics of possibilities—and theories of knowledge about how to actualise these imagined possibilities.

This will require a deepening in our understanding of how social movements can be recognised as mobilising knowledge and resources with the aim of generating tipping points in political values, actions and behaviours and helping us avert, or adapt to, potentially dramatic tipping points in human/ecological systems. In order to begin this process, I would argue we need to refine our understanding of how we undertake social movement research and to question the meaning of social movements for the academy.

Ethically informed and participatory approaches to researching with marginalised or oppressed communities are familiar from participatory action research (PAR); Friere, 1972; Chambers, 1983), which explicitly incorporates an ethic of inclusion and 'care' (Gilligan, 1982) and also from within critical approaches to qualitative inquiry (Denzin et al., 2008). I will argue that a further ethical consideration here is to conduct research that is consistent with the idea, voiced by social movements themselves (Colectivo Situaciones, 2005; Graeber & Shukaitis, 2007) and key movement theorists (Melucci, 1989, 1996b), that social movements are spaces of knowledge production about the limits and possibilities of agency and structure within a given society. I will further argue that Melucci's (1989, 1996b) ontology of movements is in keeping with a realist and materialist orientation in social science and that this provides the most appropriate conceptual framework for engaging with the social and political ontologies and epistemologies produced by social movements.

Engaging with Movements: Ontology First?

Social movement activism is normally premised on a critique of the status quo and an analysis of possibilities for change (Gillan, 2008). It requires the communication of both critiques and alternatives and an appeal to what is presumed to be the shared morality of wider publics, for whom the alternatives asserted might be regarded as superior on grounds of equality, justice or sustainability. In this sense, movements operationalise the idea of the ontological as a realm of possibility—a strategy illustrated by the slogan 'Another World is Possible' (Fisher & Ponniah, 2003), which subsequently became pluralised to 'Other Worlds Are Possible.' Ontology comes first, the assertion of a political imaginary that is recursively developed epistemologically through experiences of experimentation with counter-normative forms of expression, relationship or lifestyle.

Consequently, the knowledge-practices of movements frequently result from embodied and affective experiences that are outside of the analytical standpoint of the academic, whose methodology is reliant upon the presumed reliability of simple representative forms—interviews and other texts that 'retain an implicit ontology of the "empirical world"' (Outhwaite, 1987, p. 32). From the perspective of critical realists, this is an 'epistemic fallacy' (Collier, 1994, p. 137) which has misled positivists, empiricists and 'rationalists' since Descartes (Outhwaite, 1987, pp. 36–37; Collier, 1994, p. 137). It is this critique and its relationship to social movement studies, methodology and ethics that we can now begin to explore.

What are We Seeking Knowledge of?

In order to know, we must first have something to know about, and we must distinguish what properties this 'something' has in order to ascertain whether it is a possible source of knowledge for us (Bhaskar, 1989, p. 13). For social scientists, the sources of possible knowledge are people and societies, and as Collier (1994) observes we are therefore obliged to offer some model of what people and societies are or might become. This, of course, is also a foundational question for social movements concerned with radical social change. Meaning that if social scientists are to act ethically towards social movements and to treat them as knowledge-producers, social scientists are required to take account of the ontological frameworks movements advance—the political imaginaries and alternative accounts of what might be possible within a given society.

In order to attempt this, this article will show how methodological guidelines for engaging in the co-production of knowledge with social movement actors can be elicited from within 'classical' attempts at articulating complex ontologies. After consideration of these perspectives, I will then turn to the social movement theorist who is perhaps most sensitive to these issues—Melucci (1989, 1996a, 1996b), and describe his journey from ontology to methodology in his analysis of collective action, replete with critical reflection on the effectiveness or indeed the desirability of such a journey. Furthermore, I will argue that a 'realist' ontology is ethical because it provides for claims to knowledge 'by' social movements that have urgent social and political implications, rather than knowledge 'about' social movements which treats their own concerns as secondary or relative to their own specific ontology/cosmology.

I begin this task by confronting the dualistic heritage of 'sociological modernism' (Stones, 1996, p. 1), the somewhat artificial division between structure and agency, which continues to problematise our understanding of what we can know. Possibly, the most well-known attempt to resolve this dichotomy is Giddens' (1984) 'Structuration Theory' and I will try and show how consideration of such frameworks might lead us deeper into consideration of an informed and ethically consistent research methodology.

Giddens' conception of society is 'relational'; that is, we are defined in a given context by our relation, one to another. My subjectivity is constituted as a product of a complex set of relations situated within a web of other relations and material contexts. Subjectivity is continuously (re)constructed and therefore continuously (re)constructs; self and society are therefore two sides of the same coin. Society is, and can only ever be, people and the material results of their actions (Collier, 1994, p. 139). However, this does not imply some reductionist or individualist conception of society; rather it demonstrates the complexity of a society, where who we are becomes a matter of who is asking the question because of the multiple networks of relationships in which we are embedded (Melucci, 1996a, p. 50).

Although Giddens' articulation of a rich, diverse and flexible ontology is premised on a relational understanding of how society and the self are formulated, and this accords with other advanced ontologies such as those of the critical realist Bhaskar (1989), it also suggests epistemological and methodological questions which ultimately neither Giddens (1990) nor Bhaskar (1989) resolves.

Giddens' own foray into methodology alights upon a reflexive and hermeneutically informed ethnographic research as consistent with his ontology. However, his apparent unwillingness to develop or advocate a methodology incorporating these musings has been interpreted by Stones (1996, p. 88) as a reluctance to engage with a view of epistemology and methodology 'taken from the past,' where 'methodological rules were mean spirited and narrow, desiring to squash all the variety of the world into the same small bottle.' Although Giddens demonstrably privileges a methodology that seeks to represent social phenomena in all their ontological richness, the ontologically sensitised researcher, seeking guidance on co-producing knowledge, is left wondering how she/he might argue that what they are engaged in is in some way different from the approach of those who do not privilege ontology before epistemology.

In seeking to avoid a prescriptive approach to methodology, thereby avoiding the critique levelled at logical positivists, Giddens is susceptible to the criticism that he has not moved beyond an 'anything goes' approach (Feyerabend, 1993) and is therefore complicit in what Stones (1996, p. 20) calls 'defeatist postmodernism.'

Therefore, when we examine a classic account of social research, such as Giddens' (1984), we find that although it is ontologically bold, it remains methodologically cautious, as Giddens chooses to avoid spelling out what he infers would be an unnecessary and prescriptive set of methodological guidelines. Rather, one is left with the feeling that Giddens' structuration theory might act as the conscience of the researcher. Whilst persuasive, his ontology appears 'arrived at,' rather than 'worked out,' or as Stones (1996, p. 32) describes it '(a) common-sensical method of constructing an ontology on the basis of selective critique.' Consequently, it is worth looking at a further 'realist' approach to the same problem.

Bhaskar and Critical Realism

One of the most influential attempts to elucidate a complex and sophisticated ontology, and that from which the term 'critical realism' is derived, is Bhaskar's (1979) work. Bhaskar's (1979, 1986, 1989) realist ontology is premised on a response to the Kantian question—how is science possible? Or, as Collier reframed it—how are experiments possible? (1994, p. 31) Outhwaite summarised the Bhaskarian reply as follows:

> ... for science to be possible or intelligible, the world must be made up of real things and structures (Outhwaite, 1987, p. 31).

Bhaskar argues that the conjunctions observed in a laboratory, which occur because of careful manipulation by scientists, do so due to the intrinsic properties of the objects involved. Consequently, we must presume the continued existence of such properties outside the laboratory, whether or not they are ever demonstrated. The 'open systems' of the social and natural world are what Bhaskar is concerned with, and he has sought to illuminate the complexity and contingency of simple acts in order to demonstrate the empiricist fallacy of presuming that regular conjunctions are the result of simple causal chains. This led Bhaskar to develop his model of the 'stratification of reality,' which, as Stones observes, has a 'three-fold ontological distinction' (1996, p. 29).

The distinction is between:

(1) the real mechanism—an object's inherent powers;
(2) the actual event—the dormancy or actualisation of those powers and
(3) the empirical event—that aspect of the event which is observed.

The ontological implications of this 'stratification of reality' undermine theories that place a different, normally inferior ontological status on causal mechanisms. Whilst mechanisms are not experienced and do not occur as events, they are nonetheless the basis of the causal criteria we routinely apply to reality, as Collier notes:

> *within* the level of the Actual we are employing causal criteria all the time, and would never get out of the Empirical if we did not: when we find the garden muddy in the morning, we assume a real rainstorm, though we slept through it; a murder victim implies a murderer, even though one might never be identified (1994, p. 44).

Less sophisticated ontologies ignore causal mechanisms, preferring only those things that are a possible object of experience, thus drastically flattening out what might be 'known'. To summarise then, the domain of the real (that which is inclusive of the multiplicity of hidden causal mechanisms) is greater than or equal to the domain of the actual (what occurs), which in turn is greater than or equal to the domain of the empirical (that which can be observed) (Collier, 1994, p. 45). In open systems, there occur a multitude of events that are jointly co-produced by a multiplicity of mechanisms, so that whilst it is possible to arrive at partial closure in the natural sciences, thereby demonstrably exhibiting a causal mechanism, this is not the case in the social sciences. Theoretically, however, because of the 'stratification of reality' and the existence of causal mechanisms, events in principle *are* explicable. Stones has taken this principle as a methodological benchmark, which, whilst evidently unachievable, serves 'as a point of reference by which we can judge how far our accounts fall short of the rich reality of events' (1996, p. 7). Methodologically, then, this principle obliges:

> ...the answering of a research question with utterly exhaustive detail; absolutely everything relevant to a question in terms of, for example, hermeneutics and contiguity or power or material resources, must be included in order for an account to be exhaustive and closed to any further extension. (1996, p. 7).

Although utopian, this ideal allows us to ponder the implications of a critical realist ontology for the research process of engaging with social movements.

However, like Giddens, Bhaskar refrains from discussion of research techniques that would concur with his ontology, leaving us to look elsewhere for methodological guidelines for engagement in the co-production of knowledge with social movement actors. Here, I turn to the study of collective action as outlined by Melucci (1989, 1996a, 1996b), which I would argue is consistent with the disposition towards ontology, epistemology and methodology outlined above.

Melucci: Reconstructing the Complexity of Collective Action

Alberto Melucci first introduced the term 'new social movements' into sociological literature in 1977 (later to be translated into English: Melucci, 1980, 1981). It is a term which, although problematic (Melucci, 1995, pp. 107–119), has come to be enormously influential in research on collective action, thereby establishing Melucci as a prominent figure in the sociology of social movements, a theorist whose work:

16

...recast existing theoretical approaches under a new focus and thus has raised innovative concepts and frameworks to explain contemporary social movements (Bartholomew & Mayer, 1992, p. 141).

There are a number of reasons for situating Melucci's work within the debate about the epistemological and methodological consequences of an ontology drawn from structuration theory and critical realism. The rapid growth in collective action around issues of race, gender, disability, age, peace and the environment led Melucci to a rejection of inherited analytical frameworks. Social movements became indicative of what Melucci termed 'frontier land' (1992), the open space which reveals the contingency of any existing order. This is the marginal space, the space revealed by analysis of structure and agency, and is therefore uniquely implicated in both Giddens' (1984) discussion of structuration theory and Bhaskar's (1989) critical realism. As Melucci observed:

this field of sociological research is like a gymnasium for students who wish to take on the challenge of investigating the relationship between social systems and actors (1992, p. 239).

Melucci (1988) begins his critique of existing analytical frameworks by criticising their limited ontology. Traditional structural-functionalist perspectives (Smelser, 1962), he argues, represented collective action as a response to a perceived disorder. Alternatively, Marxists were apt to conceive collective action as an expression of the objective conditions arising from a capitalist mode of production, which also lent meaning to the action. Melucci (1988, p. 335), however, rejects both perspectives as suffering from the same 'epistemological misunderstanding'. Their error, he argues, is the treatment of a collective phenomenon as a *'unitary empirical datum'* (1988, p. 330 original emphasis):

The occurrence of certain concomitant individual behaviours forms a unitary *gestalt* that is transferred from the phenomenological to the conceptual level and acquires ontological consistency: The collective reality exists as a thing (1988, p. 330).

The two 'traditional' models, therefore, have a series of ontological building blocks from which to theorise. The causal factors—motivations, anxieties, debate, communication, discussion and negotiation—which are central to the maintenance of any collective identity, and more particularly any collective action, are already lost. The ontologies of the Marxist and the functionalist are drastically flattened, as Bartholomew and Mayer (1992, p. 142) point out when citing Melucci (1989): 'The result is a view of social movements marching through history "towards a destiny of liberation, or as crowds in the grip of suggestion"'. This criticism of the flattening of ontology to the empirical through a teleological perspective, or what Melucci referred to as the 'myopia of the visible' (1988, p. 337) is of course reminiscent of the critique Bhaskar (1979, 1989) mounted against positivism; such a flattened ontology is unable to elucidate causal factors in any other than the most reductionist terms.

Melucci also suggests that the significance of the term 'new social movement' has become its contribution to the same reductionist conceptions of collective action. The argument as to whether something is 'new' or not is of course a relative one that Melucci originally deployed to demonstrate the weakness of the existing theoretical standpoints. Defender and protagonist alike make the same epistemological error when referring to the 'new social movements' in debates as to whether or not they are new. Invariably, the debate is grounded in an understanding of a movement as a unified empirical object and thus:

> they fail to recognize that collective action always consists of various components (analytical levels, types of relationship, orientations and meanings) (Melucci, 1995, p. 110).

Melucci's emphasis upon causal mechanisms, interaction of praxis and structure, and the continuing construction and reconstruction of a movement, whose contingency is framed by such processes, accords with the ontological perspectives of both Giddens and Bhaskar. In the following extract, which bears citing at length, Melucci indicates with precision the epistemological implications of an ontology which sits easily with both Giddens' (1984) structuration theory and Bhaskar's (1989) critical realism:

> Collective action should thus be considered as the result of purposes, resources and limits: as a purposive orientation constructed by means of social relationships within a system of opportunities and constraints. It therefore cannot be viewed as the simple effect of structural preconditions or the expression of values and beliefs. Individuals acting together construct their action by means of organized investments: that is, they define in cognitive, affective and relational terms the field of possibilities and limits which they perceive, while at the same time activating their relationships so as to give sense to their being together and the goals that they pursue (Melucci, 1995, p. 111).

Melucci acknowledges the explicit relationship between his ontology and that articulated by Giddens, citing Giddens (1984) as a necessary brake upon a radical constructivism in which it is impossible to sustain when one recognises that: 'Action is an interactive, constructive process within a field of possibilities and limitations recognised by actors' (1992, p. 254). Given the ontological parallels between Melucci and what has been characterised here as a realist ontology, it is perhaps unsurprising to find that the epistemological consequences of Melucci's ontology lead to his advocating a methodological framework which he believes is consistent with his philosophical insight:

> It is not only necessary to develop new techniques, but also to make ever more explicit, as part of the research process itself, the social relations and options which are the basis for the practices adopted and that make it possible (1992, p. 256).

This journey from ontology through epistemology to methodology accords with that advocated by Stones (1996, p. 19), whose emphasis upon the 'social agent's frames of meaning' and the context-dependency of those meanings parallel Melucci's (1992, p. 256) 'situational epistemology,' both emphasise the importance of mediations, and both seek to uncover the mechanisms by which social phenomena are produced. In this sense, it is clear that Melucci adheres to Stones' concept of a 'sophisticated realism'—'whose acknowledgement of a rich and complex ontology is accompanied and matched by the adoption of a finely grained set of reflexive guidelines' (Stones, 1996, p. 232). This leads us to the consideration of how this is enacted and where we might look for evidence of this in practice. It might also be unsurprising that such examples often occur outside the academy.

The Practice of Knowledge Production

What we must recognize is that actors themselves can make sense out of what they are doing, autonomously of any evangelical or manipulative interventions of the researcher. And in the disenchanted world of consummate systemic processes where epistemological privileges have been divested together with everything hereditary and natural, all meaning is judged not by the correctness of its content but by the processes of its creation. (Melucci, 1996b, p. 389)

This article argues that one of the central ethical issues in researching social movements is a failure to make explicit the ontological and epistemological premises that underpin social movement research. This I suggest can lead to positions where movements are overly determined as 'unitary empirical datum' (Melucci, 1996a, 1996b, p. 330) or alternatively they are interpreted largely within their own frames of reference and without challenge, whereas attempts at co-production require more substantial discussion on the foundations of knowledge claims, which inevitably requires reflection on the ontologies and epistemologies of all parties. Put simply, the contention of the paper is that the social and political ontologies and epistemological practices of contemporary social movements should be taken seriously if one is to act ethically in relation to these movements.

Social movements have long been bearers of knowledge about forms of oppression and injustice, expressing political claims, identifying social and economic grievances and bringing new or neglected issues to public prominence. They have been in the forefront of debates about how social divisions, including gender, race, sexuality, age and religion, structure society and reproduce power structures including prevailing norms and values, as well as debates about the possibilities of agency in social change processes. They have also been prominent in highlighting the social and environmental implications of the application of new sciences and technologies from manufacturing processes to nuclear fission, genetically modified organisms to cloning and nanotechnology.[2] Social movements produce knowledge that is often challenging to those in power or which might be difficult for a society to confront—levels of sexual abuse, the treatment of the mentally ill,[3] the stigmatisation of those with HIV/AIDs, etc. However, rarely are social movements explicitly recognised as producers of knowledge, despite their influence in shaping various academic disciplines including womens' studies, peace studies, adult and popular education, black and post-colonial studies, queer studies, etc. These disciplines have been richly innovative in both their approaches to, and work with movements, and yet for the most part social movements are still considered as objects of knowledge for researchers and academics, rather than as knowledge producers in their own right.

This began to change during the final decades of the 20th century and the first decade of the 21st century, during a wave of social movement mobilisation focussed on resistance to the globalisation of neo-liberal capitalism and the promotion of 'social justice' (Notes from Nowhere, 2003). It can be argued that these movements were prescient in identifying issues that would subsequently move to the centre of mainstream political debate and were amongst the first to offer a cogent critique of the inequity and structural limitations of the global financial system that had been emerging from the late 1980s (Dannaher & Burbach, 2000; Houtart & Polet, 2001). It has also been argued that the reflexive practices of social movement activism during this period, from the Zapatista inspired encuentros (Holloway & Pelaez, 1998) to the dialogical and deliberative spaces of the World Social Forum (Sen *et al.*, 2004), represented a qualitative shift in the methodology of global social movements. This is important because meaning making and knowledge production subsequently became key activities of these movements (Cox & Fominaya, 2009).

This wave of mobilisation produced a generation of academic-activists and activist-researchers who have sought to challenge the epistemological premises of orthodox social movement studies. Including theoretically and methodologically operationalising long-established critiques of positivist and Cartesian epistemologies, by blurring the boundaries between the subject and object of knowledge and pursuing practices of co-producing knowledge with, rather than on movements (Conway, 2004; Casas-Cortés *et al.*, 2008; Cox & Fominaya, 2009). This milieu has produced some fascinating work on the specificities of local and indigenous knowledges and their implications for understanding the 'global' (Escobar, 1998, 2009), as well as a variety of attempts to develop 'knowledge-practices' (Casas-Cortés *et al.*, 2008) that bridge academic and movement domains (Notes from Nowhere, 2003; Sen *et al.*, 2004; Colectivo Situaciones, 2005; Graeber & Shukaitis, 2007; Turbulence Collective, 2010).

The compound term 'knowledge-practice' has much in common with the older but closely allied concept of cognitive-practice (Eyerman & Jamison, 1991) and it reflects theorisations of social movements as knowledge producers, rather than merely as objects of knowledge for social movement scholars. The use of the compound term 'knowledge-practice' (Casas-Cortés *et al.*, 2008) is also used to indicate where activism is understood as productive of critical subjectivities whose situated and contextual knowledge is prioritised in its own right. The danger being here, of course, is that this simply becomes a 'relativist' or 'anything goes' approach, where valorising the contextual undermines any claims to the wider value or application of the knowledge being produced.

So what methodological insights are to be derived from the complex and 'realist' ontology outlined above and how might these reflect the many situated 'truths' discovered by social movement experimentation with new forms of social and political life. Qualitative engagement through participatory methodologies ranging from PAR (Friere, 1972; Chambers, 1983) to visual and cartographic methodologies[4] to autoethnography (Reed-Danahay, 1997; Ellis, 2008) and autoethnographic activism (BRE, 2007) offers insights into a practice of co-production which are sensitive to a situated epistemology. Consequently, I continue to argue that developing this bricolage of approaches is akin to '... an ethical attempt to humanize activist and academic practice—to consider human bodies, desires, endurance, affects, quirks—to create an new activist and intellectual ethic.' One that is 'based on the idea that self-making and ethics are at the core of any effective and radical political project.' (Osterweil & Chesters, 2007, p. 254).

However, such methods are often described as invoking a set of 'dangers' prescribed by the academy because they constitute a tendency towards 'surrender' or 'becoming' (Hammersley & Atkinson, 1995, p. 115). So, whilst mainstream ethnography warns of the need to critically evaluate the role of the researcher towards the researched, and assert the need to suspend judgement upon the supposed reality of appearances, there remains a firmly held belief in the need to maintain a critical distance between researcher and researched (Evans, 1992, pp. 200–01). This also involves establishing clear boundaries as to what constitutes the research context, which Hammersley and Atkinson (1995, p. 112) assert requires the ethnographer to be 'intellectually poised between familiarity and strangeness; and, in overt participant observation, socially he or she will usually be poised between stranger and friend'. This leads them to suggest that ethnographers 'must strenuously avoid "feeling at home"' (1995, p. 115) and that the researcher must remain a 'marginal native' (1995, p. 112).

However, these arguments appear deeply problematic in the light of the work of those concerning themselves with the practice of co-producing knowledge with social movements (Conway, 2004; Casas-Cortés *et al.*, 2008; Cox & Fominaya, 2009). They also contrast with the potential of a critically reflexive and emancipatory research practice, such as that identified in feminist epistemologies (Harding, 1987; Lather, 1991; Stanley & Wise, 1993; Gibson-Graham, 2006), as well as those identified by anthropologists (Cohen, 1992; Escobar, 1998).

Melucci (1996b, pp. 393–397) has also been critical of the implicit assumptions of such an approach, which often remain unchallenged:

> Acknowledging both in ourselves as scientists and in the collective actors the limited rationality which characterizes social action, researchers can no longer apply the criteria of truth or morality defended *a priori*, outside of the relationship. Researchers must also participate in the uncertainty, testing the limits of their instruments and of their ethical values. (1996b, p. 395)

For Lather, it is acknowledgement of this uncertainty that characterises emancipatory feminist research practice: 'courage to think and act within an uncertain framework emerges as the hallmark of liberatory praxis in a time marked by the dissolution of authoritative foundations of knowledge' (1991, p. 13).

A process of ethical research, then, considered from these perspectives, is an unfolding of obligations and limitations developing from the relational dimension of the interaction. This requires one's own position of power, security or vulnerability to be open to analysis and contest. The academy has no *a priori* reason or justification for making demands upon those it seeks knowledge of; indeed, if we are serious about countenancing an ethical research practice, it is necessary to situate that practice within an explicit description of how the research is sustained and from where our sources our support are drawn.

There is often an implicit expectation amongst social activists that academics researching with social movements will intervene to offer resources and/or expertise during encounters with the state or the corporate sector and this is sometimes romantically presented by academics as risk-taking behaviour. However, such narratives do a disservice to the relational obligations of reciprocity derived from activist support for research processes, by providing access to information and resources that they have no prior obligation to give. Consequently, although engagement within supportive actions is associated in the academic literature with the risk of compromising the 'goal of producing knowledge' (Hammersley & Atkinson, 1995, p. 286), they are instead, invaluable articulations of the ethic of co-production, which recognises the epistemological complexity of respecting movement ontologies. Hammersley and Atkinson's definition of ethnography would therefore appear at odds with the methods one might derive from a Meluccian approach to the study of movements:

There must always remain some part held back, some social and intellectual distance. For it is in the space created by this distance that the analytical work gets done (Hammersley & Atkinson, 1995, p. 115).

I believe this artificial separation between analysis and action is merely a remnant of the hidden positivism that other ethnographers have warned against (Evans, 1992). Whilst Hammersley and Atkinson acknowledge the practical and emotional difficulties some ethnographers have had when 'leaving the field' (1996, p. 121) and the frequent establishment of friendships which persist 'for a long time' (1996, p. 122), they retain an explicit belief in the 'identity' of *the researcher* and consider other interventionary actions as either not constituting or even inhibiting the production of knowledge (1996, p. 286). This has led to criticism of their role in excluding sources of 'Other' knowledge:

> Writers such as Hammersley (1995), exercise that exclusion by declaring feminist, anti-racist, critical and emancipatory 'truths' outside the norms of legitimate research. By a discourse of derision they are dismissed as prejudiced, ignorant and ideological. In doing so the threat to notions of knowledge and to sources of income, is diverted. We are not talking about different kinds of knowledge of equal status. Stanley (1990, p. 5) describes how, within the 'academic mode of production' official and unofficial gatekeepers use myriad ways of controlling academic inputs and outputs. At the centre of these is a notion of scientism, grounded in Cartesian dualisms as to who can be a knower and what can be known, and concerned with producing knowledge through the observation of the real—those objects which exist independently of our beliefs about them. It explicitly excludes knowledge produced through alternative research approaches. (Humphries, 1997, p. 4.6)

Melucci, however, attempts to hold these tensions in balance by acknowledging the delimiting characteristics of the role of 'researcher', whilst accepting the multiple levels of engagement that any person might have within the research context.

> The point at which the two (researcher and subject) can meet can only be contractual in nature. There is nothing of the missionary about this, and the contractual meeting allows no researcher expectations as to the destinies of the actors. The researchers may be involved as individuals, as citizens, as political militants, but not as specialists. As such, they have the task of performing a professional role within knowledge producing institutions. They are therefore the bearers of the ethical and political responsibility for the production and allocation of cognitive resources; but they do not have the right to orient the destinies of society as 'counsellors of the Prince' or as ideologues of protest (1996b, p. 391).

This position exhibits a creditable acknowledgement of the inherent tensions in maintaining and negotiating a coherent identity as a researcher (academic/activist) in a dynamically shifting complex of social relationships; however, it could also be prescriptive and limiting for those activist researchers whose situated epistemology is derived from outside the limitations of a professional role as a knowledge producer. This is also, of course, an ethical dilemma for those activist-academics within the University who, whilst accepting the ethical and political responsibility for the construction of knowledge within an academic context, retain lay knowledge as either local residents, citizens or activists. As well as experiencing the emotive and affective bonds with fellow activists, friends and neighbours who inevitably shape their wishes for the outcome(s) of the event, action or campaign and the destinies of the actors involved.

Conclusion: Advocating a Meluccian Methodology in the Co-Production of Knowledge

Melucci, I would argue, through his adherence to what Stones' (1996) terms 'sophisticated realism', made a significant advance in our understanding of how to research social movements and collective action, even if those insights have yet to be thoroughly understood and taken up. I have sought to demonstrate that Melucci's approach to social movement studies is in accordance with a complex and realist ontology, and that such an ontology obliges an epistemological position which in turn suggests a particular approach to research. Melucci, by explicitly addressing the question of how one might research social movements (1992, 1996b), has refused to duck the question so adeptly sidestepped by those such as Giddens (1989).

Whilst accepting that Giddens' (1984) 'structuration theory' overcomes the structure/agency dualism which has bedevilled much of sociological inquiry since the 19th century, I would concur with Stones, who asserts that: 'Giddens' disinterest in epistemological matters and his loose and unsystematic attitude to methodology means that, in practice, he has few rules of sociological research' (1996, p. 92). There appears in Giddens' theory, I would suggest, a reluctance to engage with methodology because of the diversity and plurality of research problems and potentialities. If, then, we are to conclude that an 'anything goes' approach is fine as long as we are 'loosely sensitised' by structuration theory, its ontological significance seems to have been vastly over-rated. Melucci (1996b), on the other hand, is quite clear about the implications his ontology has for the research process. The dualistic tradition of 'density of structures versus mobility of actors' has indelibly marked the 'conceptual models and research practices in the field of collective action' (1996b, pp. 381–82). Therefore, it is imperative that new, suitably reflexive and hermeneutically inspired research procedures are developed in order to be ontologically consistent. To do otherwise is to risk repeating traditional methodological assumptions that have by inference flattened out the bold ontological terrain we have already charted. Method cannot be separated from ontology, and ontology has epistemological consequences, as Melucci asserts towards the end of *Challenging Codes*:

> This is not a question of innovative techniques alone. It entails, as part of the very process of research, rendering ever more explicit the social relations and the options that provide the procedure with its basis and which make it possible. In other words, what is called for is, as it were, a situational epistemology, which social research increasingly needs if it is to break out of the illusion that it stands outside or above the circular observer-actor game (1996b, p. 396).

Then perhaps, we may, through a systematically theorised methodology, get close enough to the 'real' to argue, as Melucci does, that: 'a limited situated knowledge can become "true" when it carries with it the awareness of its own limitations' (1996b, p. 396).

Indeed, such 'truths' may well be the source of knowledge-practices that can begin to address the convergence of complex crises in the early 20th century.

During the course of this article I have argued that the widespread acknowledgement of social movements as producers of knowledge is not matched by a commitment to social movement research which takes seriously an ethical responsibility to respect the ontological and epistemological frameworks of knowledge production that emerge in social movements. I have further argued that in order to begin the task of co-producing knowledge, we must reflect upon the ontological and epistemological frameworks that inform the academy's engagement with movement actors in knowledge production, and I have drawn attention to the importance of social movements as actors exploring the possibilities and limitations of structure and agency. Underpinning this endeavour is the normative presumption that social movements are interesting because they produce knowledge of the possibilities and limitations of human societies, knowledge which the social sciences needs to urgently engage with and contribute to, if more creative and desirable means of attending to complex problems are to be developed.

Notes

1. This is a term kindly suggested to me by Kevin Gillan.
2. For examples see, respectively: Clamshell Alliance (http://www.clamshell-tvs.org/), Say No to GMOs (http://www.saynotogmos.org/) and The International Disability and Human Rights Network (http://www. daa.org.uk/index.php?page = left-bioethics).
3. See Mad Pride (http://madpride.org.uk/index.php).
4. See Counter Cartographies Collective (http://www.countercartographies.org/).

References

Bartholomew, A. & Mayer, M. (1992) Nomads of the present: Melucci's contribution to 'New Social Movement Theory', *Theory Culture and Society*, 9, pp. 141–159.

Bhaskar, R. (1979) *The Possibility of Naturalism* (Brighton: Harvester Press).

Bhaskar, R. (1986) *Scientific Realism and Human Emancipation* (London: Verso).

Bhaskar, R. (1989) *Reclaiming Reality* (London: Verso).

BRE (2007) 'Hard Livin': Bare life, autoethnography and the homeless body, in: D. Graeber & S. Shukaitis (Eds) *Constituent Imaginations: Militant Investigation, Collective Theorization* (Edinburgh: AK Press).

Casas-Cortes, M., Osterweil, M. & Powell, D. E. (2008) Blurring boundaries: Recognizing knowledge-practices in the study of social movements, *Anthropological Quarterly*, 81, pp. 17–58.

Chambers, R. (1983) *Rural Development: Putting the Last First* (London: Longman).

Chesters, G. & Welsh, I. (2006) *Complexity and Social Movements: Multitudes at the Edge of Chaos* (London: Routledge).

Cohen, A. (1992) Self-conscious anthropology, in: J. Okely & H. Callaway (Eds) *Anthropology and Autobiography* (London: Routledge).

Colectivo Situaciones (2005) Something more on research militancy: Footnotes on procedures and (in)decisions, *Ephemera*, 5(4), pp. 602–614.

Collier, A. (1994) *Critical Realism: An Introduction to Roy Bhaskar's Philosophy* (London: Verso).

Conway, J. (2004) *Identity, Place, Knowledge: Social Movements Contesting Globalization* (Halifax: Fernwood Publishing).

Cox, L. & Fominaya, C. F. (2009) Movement knowledge: What do we know, how do we create knowledge and what do we do with it?, *Interface: A Journal For and About Social Movements*, 1(1), pp. 1–20.

Dannaher, K. & Burbach, R. (Eds) (2000) *Globalize This!* (Maine: Common Courage Press).

Denzin, N., Lincoln, Y. & Tuhiwai Smith, L. (Eds) (2008) *Handbook of Critical and Indigenous Methodologies* (London: Sage).

Ellis, C. (2008) *Revision: Autoethnographic Reflections on Life and Work* (Walnut Creak: Left Coast Press).

Escobar, A. (1998) Whose knowledge? Whose nature? Biodiversity, conservation and the political ecology of social movements, *Journal of Political Ecology*, 5, pp. 53–82.

Escobar, A. (2009) *Territories of Difference: Place, Movements, Life, Redes* (Durham and London: Duke University Press).

Evans, M. (1992) Participant observation—The researcher as research tool, in: J. Eyles & D. Smith (Eds) *Qualitative Methods in Human Geography*, pp. 17–38 (Cambridge Polity Press).

Eyerman, R. & Jamison, A. (1991) *Social Movements: A Cognitive Approach* (Cambridge: Polity Press).

Federici, F. (2009) Education and the enclosure of knowledge in the Global University, *ACME: An International E-Journal for Critical Geographies*, 8(3), pp. 454. Available at: http://www.acme-journal.org/vol8/Federici09.pdf.

Feyerabend, P. (1993) *Against Method* (London: Verso).

Fisher, W. F. & Ponniah, T. (Eds) (2003) *Another World Is Possible: Popular Alternatives to Globalization at the World Social Forum* (London: Zed).

Friere, P. (1972) *Pedagogy of the Oppressed* (London: Penguin).

Gibson-Graham, J.K. (2006) *A Postcapitalist Politics* (Minneapolis: University of Minnesota Press).

Giddens, A. (1984) *The Constitution of Society: Outline of the Theory of Structuration* (Cambridge: Polity Press).

Giddens, A. (1989) A reply to my critics, in: D. Held & J. Thompson (Eds) *Social Theory and Modern Societies: Anthony Giddens and his Critics* (Cambridge: Cambridge University Press).

Giddens, A. (1990) Structuration theory and sociological analysis, in: J. Clark, C. Modgil & S. Modgil (Eds) *Anthony Giddens: Consensus and Controversy* (London: Falmer).

Gillan, K. (2008) Understanding meaning in movements: A hermeneutic approach to frames and ideologies, *Social Movement Studies*, 7(3), pp. 247–263.

Gilligan, C. (1982) New maps of development: New visions of maturity, *American Journal of Orthopsychiatry*, 52(2), pp. 199–212.

Graeber, D. & Shukaitis, S. (2007) *Constituent Imaginations: Militant Investigation, Collective Theorization* (Edinburgh: AK Press).

Hammersley, M. (1995) *The Politics of Social Research* (London: Sage).

Hammersley, M. & Atkinson, P. (1995) *Ethnography: Principles in Practice* (London: Routledge).

Harding, S. (Ed.) (1987) *Feminism and Methodology* (London: Open University Press).

Harvey, D. (2010) *The Enigma of Capital* (London: Profile Books).

Holloway, J. & Pelaez, E. (Eds) (1998) *Zapatista! Reinventing Revolution in Mexico* (London: Pluto Press).

Houtart, F. & Polet, F. (Eds) (2001) *The Other Davos: The Globalization of Resistance to the World Economic System* (London: Zed).

Humphries, B. (1997) From critical thought to emancipatory action: Contradictory research goals? *Sociological Research Online*, 2 (1).

Johnson, S. (2002) *Emergence: The Connected Lives of Ants, Brains, Cities and Software* (London: Penguin).

Lather, P. (1991) *Getting Smart: Feminist Research and Pedagogy within the Postmodern* (London: Routledge).

Mattick, P. (2011) *Business as Usual: The Economic Crisis and the Failure of Capitalism* (Chicago: Reaction Books).

Melucci, A. (1980) The new social movements: A theoretical approach, *Social Science Information*, 19(2), pp. 199–226.

Melucci, A. (1981) Ten hypothesis for the analysis of new movements, in: D. Pinto (Ed.) *Contemporary Italian Sociology* (New York: CUP).

Melucci, A. (1988) Getting involved: Identity and mobilization in social movements, *International Social Movement Research*, 1, pp. 329–348.

Melucci, A. (1989) *Nomads of the Present* (London: Hutchinson).

Melucci, A. (1992) Frontier land—Collective action between actors and systems, in: M. Diani & R. Eyerman (Eds) *Studying Collective Action* (London: Sage).

Melucci, A. (1995) The new social movements revisited: Reflections on a sociological misunderstanding, in: L. Maheu (Ed.) *Social Movements and Social Classes: The Future of Collective Action* (London: Sage).

Melucci, A. (1996a) *The Playing Self: Person and Meaning in the Planetary Society* (Cambridge: Cambridge University Press).

Melucci, A. (1996b) *Challenging Codes: Collective Action in the Information Age* (Cambridge: Cambridge University Press).

Meyer, D. & Tarrow, S. (1998) *The Social Movement Society: Contentious Politics for a New Century* (Lanham: Rowman and Littlefield).

Notes from Nowhere (2003) *We are Everywhere* (London: Verso).

Osterweil, M. & Chesters, G. (2007) Global uprisings: Towards a politics of the artisan, in: D. Graeber & S. Shukaitis (Eds) *Constituent Imagination: Militant Investigations Collective Theorization* (Oakland: AK Press).

Outhwaite, W. (1987) *New Philosophies of Social Science: Realism, Hermeneutics and Critical Theory* (London: Macmillan).

Reed-Danahay, D. (1997) *Auto/Ethnography: Rewriting the Self and the Social* (Oxford: Berg).

Santos, B. D. S. (2003) The popular university of social movements: To educate activists and leaders of social movements, as well as social scientists, scholars and artists concerned with progressive social transformation, Proposal for discussion. Available at: http://www.ces.fe.uc.pt/universidadepopular/Popular%20University%20of%20the%20Social%20Movements.pdf

Sen, J., Anand, A., Escobar, E. & Waterman, P. (Eds) (2004) *World Social Forum: Challenging Empires* (New Delhi: The Viveka Foundation).

Smelser, N. (1962) *Theory of Collective Behaviour* (New York: The Free Press).

Stanley, L. (Ed.) (1990) *Feminist Praxis* (London: Routledge).

Stanley, L. & Wise, S. (1993) *Breaking Out Again: Feminist Ontology and Epistemology* (London: Routledge).

Stones, R. (1996) *Sociological Reasoning: Towards a Past-Modern Sociology* (London: Macmillan).

Turbulence Collective (2010) *What Would it Mean to Win* (Oakland: PM Press).

Urry, J. (2003) *Global Complexity* (London: Routledge).

Reflexive Research Ethics for Environmental Health and Justice: Academics and Movement Building

ALISSA CORDNER*, DAVID CIPLET*, PHIL BROWN* &
RACHEL MORELLO-FROSCH**

*Sociology Department, Brown University, Providence, RI, USA, **Department of Environmental Science, Policy and Management and School of Public Health, University of California, Berkeley, CA, USA

ABSTRACT *Community-engaged research on environmental problems has reshaped researcher–participant relationships, academic-community interaction and the role of community partners in human subjects protection and ethical oversight. We draw on our own and others' research collaborations with environmental health and justice social movement organizations to discuss the ethical concerns that emerge in community-engaged research. In this paper, we introduce the concept of* reflexive research ethics*: ethical guidelines and decision-making principles that depend on continual reflexivity concerning the relationships between researchers and participants. Seeing ethics in this way can help scientists conduct research that simultaneously achieves a high level of professional conduct and protects the rights, well-being and autonomy of both researchers and the multiple publics affected by research. We highlight our research with community-based organizations in Massachusetts, California and Alaska, and discuss the potential impacts of the community or social movement on the research process and the potential impacts of research on community or social movement goals. We conclude by discussing ways in which the ethical concerns that surface in community-engaged research have led to advances in ethical research practices. This type of work raises ethical questions whose answers are broadly relevant for social movement, environmental and public health scholars.*

Introduction

For decades, academics in the USA have simultaneously studied and supported the antitoxic and environmental justice social movements. Community-engaged research on environmental problems, which also supports these movements' policy objectives, has reshaped the research enterprise in terms of researcher–participant relationships, academic-community interaction and the role of community partners in human subjects' protection and ethical oversight.

In this paper, we draw on our experiences conducting community-engaged sociological and public health research with environmental justice and antitoxic social movements to develop an expanded conception of research ethics as *reflexive research ethics*. Using this concept, we move ethics beyond formal guidelines and pre-established professional codes of conduct to consider the relational nature of social movement research and complex questions of reciprocity, control over and access to research data and research integrity. We describe the environmental justice and antitoxic movements, highlighting connections between academic researchers and their community partners, and provide an overview of community-based participatory research (CBPR) as one form of community-engaged research. We argue that doing CBPR presents new moments of ethical uncertainty, which cannot always be resolved through adherence to existing research practice, and thus require a commitment to reflexive research ethics. We discuss ethical concerns in community-engaged research related to the bi-directional impacts of academic research and social movement activity, drawing on our research with community-based organizations in Massachusetts, California and Alaska. We conclude with a discussion of how this reflexive and relational approach to research ethics can impact and improve the institutionalization of research ethics. Community-engaged research has contributed to advances in research ethics, which are broadly relevant for social movement scholars, social scientists and environmental scientists.

Social Movement Research and Research Ethics

Research ethics are often understood as 'a rather static set of standards for conduct based on a system of moral values' (Smith-Doerr, 2006), operationalized through professional codes of conduct and formalized through the institutional review board (IRB) protocols that outline acceptable research practices. IRBs are centered around the Belmont principles of ethics, including respect for persons, autonomy, beneficence and justice, which are central to community-engaged research. Belmont ethics provide valuable guidelines and minimum standards of respect for and responsibility to human subjects for all researchers. But while formalized ethical protocols provide clear and structured sets of guidelines for research, they are unable to fully take into account the dynamic relationships between multiple parties or the constant uncertainty faced by researchers as they navigate new ethical terrain or face questions unaddressed by existing standards.

While ethical uncertainty undoubtedly exists in all types of social science research, we have found that uncertainty is especially pervasive and salient during research that engages directly with communities and social movements. Ethical uncertainty is particularly acute in situations such as those that involve emerging technologies or research practices; power relationships between researchers and communities; norms that vary across communities; research benefits that are distributed across multiple communities and conflicting individual, professional and community goals. We have found that these uncertainties are not fully addressed by formal ethical guidelines, and are not necessarily resolved by adhering to them. Indeed, ethical considerations are highly connected to uncertainty, because if there was certainty about the nature of situations, we would likely not have ethical issues (Swauger, 2011).

This raises the question: without a formal ethical roadmap to follow in these moments of uncertainty, how should researchers engaged with communities and social movements proceed in an ethical manner? We take inspiration from the widespread call in sociology for reflexive and relational research (Giddens, 1984; Bourdieu, 1992; Emirbayer, 1997), and use this approach to re-conceive ethical research conduct. Reflexivity has taken on different and sometimes contradictory meanings. For example, reflexivity is understood as a process of modernization that includes self-conscious monitoring (Giddens, 1991) and individualization (Beck, 1992); a necessary methodological intervention about one's role as a researcher (Pillow, 2003; Kenway & McLeod, 2004); a practice of considering power in the construction of knowledge (Callero, 2003) and an interrogation of academia's intellectual bias (Bourdieu, 2000; Schirato & Webb, 2003). But while reflexivity has become a major theoretical and methodological influence in the social sciences, the relationship between reflexivity and ethical research conduct has yet to be sufficiently elaborated.

We introduce the concept of *reflexive research ethics* as the self-conscious, interactive and iterative reflection upon researchers' relationships with research participants, relevant communities and principles of professional and scientific conduct. We conceive of ethical conduct in research as that which maximizes beneficence and minimizes harm. Beyond this general understanding of ethics, reflexive research ethics does not assert a set of preconceived principles of how a particular moment of ethical uncertainty should be addressed; rather, it entails the continued adjustment of research practice according to more relational and reflexive understandings of what might be beneficent or harmful. Our conception of reflexivity thus fits with concerns about the partiality of perspectives in the research process; the effects of researchers on society and the nature of structural and spatial power relations between the researchers and the researched. Furthermore, we view reflexivity as not solely an individual endeavor, but rather as a collective relationship between all actors in the research process.

Conducting research with environmental health and justice organizations has shown us how ethical research is relational, in that it involves connections and exchanges between multiple parties. This type of research demands researchers' awareness of their relationships with research participants, institutions, professional disciplines and the larger community and consumers of the knowledge generated. Thus, through reflexive research ethics, we propose that when conducting research involving the daily lives and concerns of people, communities and social movement organizations, researchers must move beyond understanding scientific ethics as static, pre-conceived standards or guidelines to seeing research as a process that is embedded in social relationships and has social consequences.

Practicing reflexive research ethics requires researchers to identify and establish interactive discussions with the full range of publics relevant to their work, including research participants, local communities, academic disciplines and people potentially impacted by research findings. Researchers must also identify norms and principles that govern their research; draw upon accumulated knowledge of how others have conceptualized, addressed and reflected upon relevant ethical issues; assess ethical tensions that may arise from the prioritization of particular interests, publics or principles and respond to emergent ethical tensions.

This concept builds on a 'post-Belmont' conception of ethics, which asserts the need for a holistic approach to conducting ethical research that balances the application of Belmont principles at both the individual and community levels (Brown *et al.*, 2010). In this approach, community harm and benefit are strenuously taken into account, as well as individual harm and benefit. Sharing information is emphasized, in contrast to IRB's frequent mode of withholding information for fear of alarming people.

The ethics of conducting social movement research are often discussed in terms of questions that arise during fieldwork or ethnographic practice, such as researchers' identities in relation to the social movement they are studying, questions of access to data and dissemination of research findings (Murphy & Dingwall, 2001; Hammersley & Atkinson, 2007; Atkinson, 2009). Our concept goes farther to suggest that reflexive research ethics should govern all phases of the research process, including the *identification* of research questions and motivation; the *engagement* with community actors, social movements, knowledge institutions and other publics; the *production* of knowledge; the *interpretation or analysis* of data; the *presentation and dissemination* of research results and the *use* of scientific knowledge. This aligns with Blee and Currier's (2011: 404) point, introducing a special issue of *Qualitative Sociology* on 'Ethics Beyond the IRB', that 'ethics are involved at every stage of research'. Reflective research ethics thus are connected to the scientific uncertainty that arises throughout the entire research process, raising context-specific, relational and value-laden ethical dilemmas for researchers across a range of disciplines and methodological orientations.

Social Movements Meet Academia

Academic disciplines that address the connection between human populations and the natural environment, such as environmental sociology and environmental health science, have long grappled with co-participation, in which scholars have been involved in movements that they are also studying. As a result, these fields have, to varying degrees, been shaped historically by their interaction with social movements and community actors, and in turn, influenced the trajectories of social movements.

Although environmental and public health concerns were long part of labor and urban social movements starting in the late nineteenth century (Gottlieb, 1993), environmental sociology grew up as a research discipline contemporaneously with the modern mainstream environmental, antitoxic, and environmental justice movements, focusing on topics such as ecological limits, economic growth and increased affluence, and environmental activism (Dunlap & Catton, 1979; Dunlap & Mertig, 1992; Cable & Cable, 1995). Similarly, the environmental justice focus within environmental sociology that emerged in the 1980s directly integrated with social movements working against issues of community contamination, particularly in the US South (Bullard, 1990; Mohai *et al.*, 2009). The environmental justice movement was galvanized by a strong connection to civil rights struggles and by empirical data generated by social scientists that revealed that polluting hazardous waste facilities were disproportionately located in African-American communities (Mohai & Bryant, 1992). More recently, as environmental justice social movements have extended to areas such as food justice and climate justice, environmental sociology has followed in its path.

Although environmental epidemiology, exposure science and environmental toxicology have not been as activist oriented as environmental sociology, some environmental health scientists have engaged with social movements and community-based struggles. In part, this was due to the development and growing legitimacy of CBPR (discussed below). There was also another stream whereby public health researchers engaged with health activism through community-based organizations and patients' groups, on issues such as AIDS, women's health reforms and reproductive rights (Epstein, 2008). This historical co-production of knowledge, principles, identities and priorities between these academic disciplines and relevant social movements, and the recognition that academic research is closely connected to societal concerns, has led researchers in these disciplines to focus on addressing ethical considerations related to research practices.

Community-Engaged Research

From its origins in typically un-funded 'participatory action research' and 1960s federal programs' requirements for public involvement, CBPR grew largely through federal funding in the 1990s. CBPR is conducted as a close, collaborative partnership between academic researchers and community organizations as partners and co-researchers. In a CBPR project, research is oriented to the needs of communities and all partners are involved at every stage of the research process. CBPR projects generate scientific knowledge, improve public trust and understanding of environmental health science, inform culturally and socially appropriate intervention methods and improve public health decisions (O'Fallon & Dearry, 2002; Minkler & Wallerstein, 2008).

CBPR has been closely connected with environmental justice, both in funding mechanisms and in preferred research arrangements by environmental justice groups. Social scientists have played a central role in the expansion and institutionalization of federally funded research on environmental justice, especially through programs funded by the National Institute of Environmental Health Sciences (O'Fallon & Dearry, 2002; Senier et al., 2008; Baron et al., 2009). Not surprisingly, professional recognition, publication opportunities and university training programs have grown in tandem with such funding (Brown et al., 2010).

Rather than following only the interests and scientific objectives of researchers, CBPR is inherently relational in that the research must address issues affecting community partners, be geared toward building community capacity and report findings using accessible language and in a manner that is respectful of community needs. CBPR researchers have often sought to correct the perceived ethical shortcomings of studying communities and social movements without providing direct benefits. In all phases of research, CBPR projects are guided by ethics of mutual respect, open communication and recognition of knowledge, expertise and resource capacities of all partners, seeing them as co-learners and co-owners of data (Israel et al., 1998).

There are other forms of community-engaged research in environmental health sciences and environmental sociology. Research can be community engaged without taking the form of a formal CBPR partnership, as in popular epidemiology (Brown & Mikkelsen, 1990), citizen-science alliances (Morello-Frosch et al., 2006) and the community outreach and engagement cores of federally funded research centers (Brown et al., 2010). Such approaches demonstrate that community-engaged research can be conducted in accordance with service and outreach goals, without the formal partnerships between researchers and community organizations that characterize CBPR. Practitioners of these approaches are similar to CBPR adherents in that they generally view their approaches as driven by the ethics of individual empowerment, community empowerment and scientific and governmental responsibility to serve the public good.

The practice of CBPR constantly presents new moments of ethical uncertainty that cannot necessarily be resolved through adherence to existing practice or fixed ethical guidelines alone, including power relationships between community groups and the academic researchers who often have the funding and initiate research projects; racial, ethnic, gender and cultural dynamics; questions of data ownership; conflicts between individualistic and community-minded conceptions of ethical responsibility; distribution of benefits from research; and the programmatic requirements of funding agencies or foundations (Brugge & Kole, 2003; Minkler, 2004; Giese-Davis, 2008; Bastida *et al.*, 2010). Scholars have also discussed the challenge of conducting community-engaged research in which ethical concerns go above and beyond those clearly governed by IRBs (Flicker *et al.*, 2007; Brown *et al.*, 2010). These multivalent power dynamics make reflexive research ethics particularly salient for community-engaged research. Thus, practicing CBPR demands the reflexive approach that is central to our concept of reflexive research ethics.

Our Research

This paper identifies ethical concerns common in community-engaged research that are broadly relevant to scholars studying social movements. We discuss two general areas of ethical tensions and uncertainties that we have encountered in community-based research: the impact of research on social movement goals and the impact of social movement or community goals on the research enterprise. We give examples of research practices that build and maintain relationships between academic and community partners, and that treat research as a series of responsibilities to communities, publics and research integrity.

The examples below are drawn from our own work and that of our colleagues. In particular, we refer to three research partnerships that have been central to our research efforts in the last 10 years, all of them in the USA. In this paper, we will not provide full background information on any of these projects, as more information on the research is available from the associated citations for each project. In Massachusetts, we have collaborated on household exposure and biomonitoring studies[1] with Silent Spring Institute, a non-profit research organization that studies the links between women's health and the environment (Altman *et al.*, 2008; Brown *et al.*, 2010). In California, we have conducted a household exposure study and a health survey with Communities for a Better Environment, a community-based environmental justice organization working in an urban industrial neighborhood (Brody *et al.*, 2009; Brown *et al.*, 2011; Cohen *et al.*, 2011). In Alaska, we have worked with Alaska Community Action on Toxics, the state's only environmental health and justice nonprofit, on environmental justice concerns related to military bases and former industrial sites, persistent chemicals, and reproductive toxics (Altman, 2006, 2008). We use examples from these cases to highlight common ethical concerns, describe our attempts at reflexive research ethics and suggest the implications these ethical concerns have for other social movement researchers.

Impacts of the Research Process on Social Movement Goals

Researchers of social movements have responsibilities to the communities and individuals they study. Researchers have to balance harms with benefits, which includes informing people of possible harms they may experience in the course of the research and responding to the needs of research participants and collaborators. The research process can contribute to social movement goals, but at other times it has the potential to hinder them.

There are myriad potential negative impacts of unethical engagement of researchers with social movements and communities. One such practice has been termed 'parachute research'—dropping in to collect data without engaging with the community, and then leaving without sharing the data and results (Costello & Zumla, 2000). In such cases, the impacted community has not been engaged in the various steps in the research process, such as the development of research questions, the analysis of findings, the compilation of data or the development of policy prescriptions. This practice also diverts a community's time, labor, energy and resources into data gathering or participation in a study, without providing perceived or accessible benefits. In addition, more severe impacts can include inciting conflict between community organizations, the use of study results to support the adoption of policy measures that are deemed unfavorable by community members, data use that has unintended consequences (such as decreasing a neighborhood's property values due to publicity of contamination), and a strained financial and administrative relationship between the research institution and the community.

Reflexive research ethics can help researchers who work in communities navigate the difficulties of conducting community-engaged research within the codified ethical guidelines required by IRBs and other formal or informal academic practices and expectations. When addressing ethical challenges it is important to incorporate contextual and relational knowledge in addition to more general understandings of responsible research practice. For example, we often work with community partners in the study design phase to clearly outline our scientific and policy objectives for a collaborative project, and then work to design a study protocol and results dissemination plan that will address mutual organizational interests. Our reflexive, iterative approach to the co-production of scientific knowledge allows our research to advance the social and environmental health science fields while simultaneously supporting our community partners' environmental health and justice goals of education, public awareness and regulatory change. Here, we discuss three ethical tensions and how we dealt with them through a reflexive approach to the ethical practice of research.

The first tension we have repeatedly encountered is that certain requirements of academic research can conflict with the goals and needs of community organizations. Sometimes social movement organizations approach researchers to collaborate on specific research projects that are relevant to scientific, regulatory or advocacy goals, which may or may not coincide with academic pressures to rapidly publish results, or with financial or disciplinary constraints on study design. However, a reflexive approach to research can often unite the goals of research partners. For example, Silent Spring Institute asked us to examine the history of lay involvement in federally funded research and services, in order to justify lay involvement in breast cancer research. The organization also sought our help in communicating with the Massachusetts Department of Public Health about the scientific, logistical and practical rationale of preserving exposure sampling data for future analyses that could have public health benefits. Another organization, Communities for a Better Environment, asked us to provide briefing papers for city officials in order to facilitate organizing against a planned oil refinery expansion adjacent to the community. The third organization, Alaska Community Action on Toxics, asked us to write and later revise a brochure on reproductive health for physicians. Some of these requests (e.g. physician brochure) were straightforward in that they did not lead to potential conflicts with our university officials. But other requests, such as a desire by community partners to store sampling data after a study finished, had more potential for conflict. In this instance, we pursued multiple rounds of protocol revision with our IRB, which expressed concerns about informed-consent procedures should researchers conduct additional and unexpected analyses in the future. In doing so, we strengthened our relationship with IRB officials and developed an improved informed-consent document useful for future research projects.

This example connects to the second tension: researchers and communities face multiple difficulties when trying to combine the reflexivity of community-engaged research with the codified ethical guidelines required by IRB protocols. Community-academic collaborations involve numerous ethical decisions that move far beyond the formal ethics encoded into an IRB protocol or professional standard of ethics. This post-Belmont approach to ethics, described above, requires that both researchers and community partners play a central role in developing study protocols that ensure the protection of study participants. For example, a colleague involved in a research partnership with a community asked us for guidance after their IRB denied a research protocol, which would have tested for chemicals in breast milk. The IRB argued that women would reduce breastfeeding if they found that their breast milk contained contaminants, but we drew on existing research on women's perceptions of biomonitoring (Wu *et al.*, 2009) to show this not to be the case.

In another example, in our household exposure and biomonitoring work we have collaboratively designed study protocols so that individual participants have access to their personal exposure results if they choose, and individual- and aggregate-level report-back protocols are fully deliberated before results dissemination. However, this form of individualized results communication is controversial: some IRBs question whether uncertainty regarding the health effects of contaminants or the lack of clear strategies to reduce exposures can cause undue worry and stress among study participants. Our research documents that study participants generally want access to their personal data, even when the implications of contaminant exposures for future health effects are not well understood (Altman *et al.*, 2008). Through our community collaborations, we have developed innovative methods for reporting back individual biomonitoring results, and have conducted follow-up interviews with participants to learn how they respond to receiving their results and how they use the information (Brody *et al.*, 2007; Adams *et al.*, 2011).

The third tension relates to communicating results to communities. Although communities should not receive individual-level results except in particular circumstances and only with the expressed consent of participants, community-level results such as average exposure levels or suspected links between pollution sources and exposure often should be communicated. Environmental health researchers have documented how individual- and community-level results can be communicated in ways that work to overcome information disparities between researchers and participants (Morello-Frosch *et al.*, 2009; Emmett & Desai, 2010). In other instances when working with Native tribes, tribal IRBs have specific requirements governing community-level data communication and ownership (Watkins *et al.*, 2009).

In our work, community report-back is coordinated through partnerships with community organizations. For example, we disseminated community-level results from an exposure study and a health survey at public forums organized by our community partner, Communities for a Better Environment. Because these forums were organized by an established community organization, they were well attended and attracted the attention of local civic leaders, meaning that our research likely reached a wider audience than it would otherwise. By closely collaborating with social movement organizations, researchers can release results to the community before publishing them in scientific journals or discussing them with the media. For example, colleagues have described community meetings that were closed to the media days or hours before scientific publications or press conferences, to ensure that the first to hear about results are those most directly affected by them. This example of timely dissemination of findings demonstrates one way of respecting the community's interest in co-directing the dissemination of study results, with while ensuring timely and productive publication for academic partners.

Overall, our understanding of ethics in research has been largely shaped through active investigation into the ways in which our research has beneficial or detrimental impacts on social movement and community partners. At several points throughout a multi-year research project, we engage in mutual assessment and evaluation (through debriefings, anonymous evaluations and short interviews) in formal and informal meetings, to learn and assess how partners experience the academic-community collaborations as they evolve and whether they meet diverse objectives, needs and goals. These ongoing reflexive exercises help to ensure that emerging ethical problems are uncovered and addressed in a timely way.

Although CBPR principles can suggest ways to advance this reflexive process, they do not automatically guarantee public representation or involvement in the development of study protocols and application of ethical principles. Researchers should recognize that community groups may not be synonymous with or fully representative of the communities in which they work or the constituencies that they purport to serve. Thus, ethical research also involves being attentive to the heterogeneity of interests, needs and demographics within a particular movement, group or community. And further, it requires researchers to be aware of which publics their research seeks to serve.

Impacts of Social Movement and Community Goals on the Research Process

Research that is attuned to social movement goals and community needs has great potential for producing relevant and useful science. But there are also potential pitfalls if the researcher is also not committed to the integrity of scientific practice, and cognizant of the reality of research constraints. This situation raises ethical challenges related to balancing community or organizational concerns with scientific standards and professional obligations.

Lay identification of research questions has long benefited science, as with worker and union identification of major workplace hazards such as asbestos, cotton dust, vinyl chloride and lead (Mayer, 2008). Much of what we know about major contaminants comes from residents who report on their exposure-related symptoms (e.g. Brown & Mikkelsen, 1990). The dramatic increase in research on asthma was largely initiated by inner-city parents and teachers who noticed its growing prevalence (Corburn, 2005). If the goal of science is to gain new knowledge to make a better society, lay-initiated science has certainly shown its value in identifying emerging issues and preventing disease.

But this research must also be ethical in conforming to professional scientific standards of producing reliable and truth-oriented knowledge, whether in social science, epidemiology or toxicology. To offer a hypothetical example, a researcher works with a community group but finds that the study results contradict community beliefs about a particular environmental health problem. The ethical challenge would be how to interpret and disseminate these results. Reflexive, community-engaged research would encourage partners to talk through these scenarios before embarking on a study to decide how they might deal with such a challenge. It also helps to highlight early on the potentially high stakes of future study results and encourages a candid discussion about whether the potential benefits of conducting such a study outweigh the possible detrimental effects on community advocacy efforts.

Researchers also have the responsibility to balance community concerns with scientific constraints. Community groups may seek answers to questions that may be difficult to analyze or assess scientifically. For example, residents in a small neighborhood may believe that they have an excess of a disease, and therefore seek an epidemiological health study. Most researchers would point to the difficulty of such a study, in terms of scientific capacity, sample size and time, and might suggest a more appropriate way to call attention to excess disease, perhaps by documenting high levels of exposure or a putative source of toxic chemicals. Embarking on a research path that researchers knew was flawed would violate professional scientific standards and would ultimately undermine the interests of community partners. Community groups usually understand this, though there are exceptions. For example, a researcher well known for supporting community groups investigated a cancer cluster in proximity to an army base and faced community anger at public hearings after presenting results that did not match with the expectations of local organizations.

In order to address these challenges, we have taken many precautions in developing rigorous research projects that still support organizing and advocacy goals. We engage in community-academic collaboration in the design and implementation of the original study questions and research protocols, as well as in the interpretation and dissemination of data. Ethically minded community-engaged research can potentially enhance scientific practice through improved research design, smoother project implementation and more accurate interpretation of findings. By using a reflexive research ethics approach, we engage in the earliest possible discussion of all these issues and create spaces, as discussed above, for active reflexivity throughout the research process. This reflexivity can be somewhat institutionalized using Community Advisory Boards to hold researchers accountable by providing feedback on both scientific practice and community collaborations. Other communities have done similar things, such as the claimants in the Fernald, Ohio suit that provides medical monitoring for residents near a nuclear weapons facility (Gerhardstein & Brown, 2005).

In spite of these potential difficulties, reflexive research ethics can improve the rigor, relevance and reach of science practice (Morello-Frosch et al., 2005, 2011). For example, improved rigor can be seen in a partnership between Alaska Community Action on Toxics and environmental health researchers on indigenous food sources on St. Lawrence Island. Long-term collaboration with residents taught researchers that samples would be most meaningful if collected from hunting grounds, which were located near a decommissioned military base (Altman, 2008). Improved relevance can be seen in how Silent Spring Institute scientists incorporated the desires of Communities for a Better Environment for volunteers to be added to the household exposure study sampling frame, and they fulfilled the community's desire to look for refinery-related chemicals in environmental sampling, as well as the endocrine disrupters of primary interest to the scientists and their breast cancer-activist partners (Brody et al., 2009). Improved reach is manifested in how Silent Spring Institute scientists developed visually appealing, accessible outreach materials that study participants and Communities for a Better Environment organizers could use on their own to disseminate results.

Discussion and Conclusion

The numerous ethical concerns that emerge when conducting research with social movements or communities are best understood as *reflexive research ethics*, a series of responsibilities that researchers have to the multiple publics which their research may impact. Thus, the reflexive researcher will not view ethical practices as static or detached from the universe in which they are developed, but rather as responsive, relational and often contextual. We developed this concept through our community-engaged work with environmental health and justice organizations, specifically inspired by moments of co-production of knowledge when academics and community partners conduct scientific research collaboratively. This entanglement of the scientific enterprise with social movement and community organizing brings to the surface and amplifies ethical tensions. Thus, our concept is particularly relevant for all social movement researchers engaged in co-production of knowledge—not necessarily environmental knowledge—with communities and organizations.

However, we would argue that this constant reflexivity raises questions that all scholars studying social movements should be attentive to, including: How can research be conducted to maximize beneficence and minimize harm at all phases of the project? To whom should researchers be accountable to in terms of their research practice and modes of conduct? And what values and principles should be prioritized in their methods of engagement? The concept of reflexive research ethics does not purport to answer these questions; rather, these are questions that can be hidden in the background of the research process, unless active reflexivity brings them to the surface.

Research integrity is often understood as a commitment to adhering to established principles of research conduct in an unbiased fashion. We challenge this notion that research must be disinterested in order to be done with integrity. Rather, we view research integrity as engaging with ethics as fluid, dynamic and value-laden guideposts that must be constantly and self-consciously reflected upon. Research with community partners has demonstrated to us that the identification of what rights individuals and communities are entitled to and which practices should be employed are contested issues, and thus demand constant reflexivity on the part of researchers.

However, we have also emphasized that this re-conception of research ethics should not be reduced to simply responding to the interests or expressed needs of individuals, communities, social movements or other actors. Rather, reflexive research ethics also require a strong commitment to conforming to appropriate professional scientific standards of producing reliable and truth-oriented knowledge. Science must be seen as credible in order to contribute to the legitimacy and successes of community partners. This requires social movement scholars to actively use, but move beyond, the forms of ethical conduct, such as individual informed consent, which can be codified in IRB protocols. IRB requirements are important for research and provide necessary safeguards for research participants, but when studying organizations, communities and groups, IRB ethics often do not go far enough. Instead, seeing ethical obligations as reflexive and relational acknowledges the inviolable importance of informed consent while also recognizing that salient ethical concerns emerge during the course of social movement research that are unavoidably contextual, relational and value laden.

Not all social movement scholars need to conduct CBPR or community-engaged work, as many research questions are more appropriately answered using other research methods or types of participant observation. However, any research on social movement or community issues can be significantly improved through attention to reflexive research ethics. We are not suggesting that community engagement is a panacea for ethical concerns. To the contrary, our research documents that community-based research often magnifies or brings to the forefront ethical issues that, in less publically engaged styles of research, may not demand such high levels of attention or make some ethical conundrums inherent to the research process. Additionally, as CBPR has become more widely accepted, it becomes a potential tool for career advancement, providing further justification for reflexivity in research as researchers should ensure that chosen research strategies and practices remain beneficial for communities. Community-engaged research on its own is rarely sufficient for addressing problems of serious concern, such as environmental contamination or toxic exposure that are often central to the missions of our community partners. However, when research actively incorporates community voices and concerns, it may provide solutions to community problems which achieve scientific and non-scientific goals (such as policy-making), and support movements for justice and democratic participation in decision-making processes.

Although the research described in this paper was based in the USA, the reflexivity we advocate might travel well to other countries, especially places where interactions between researchers and participants are marked by complex and uneven relationships of power. Additionally, the questions raised by reflexive research practices about issues such as data access and dissemination, research goals and utility and human subjects protection are relevant for many types of research, including social movement studies, environmental sciences and public health. Reflexive ethics are especially important for social movement scholars studying movements whose activities deal with high levels of uncertainty, because the ethical guidelines are unlikely to be written. For example, many environmental health researchers and organizations today conduct cutting edge science that is so new that there is no consensus on what practices are ethical. Engaging around issues of scientific uncertainty requires scholars to work closely with social movement organizations to balance movement goals, research needs and the protection of individual- and community-level rights. Future research should be directed to uncovering and responding ethically to the tensions that exist in the study of emergent science and its relationship to the concerns of social movements and communities that may be impacted by such scientific endeavors.

We have described how reflexive engagement with research ethics has become increasingly institutionalized by federal funding programs that support CBPR. We suggest that the National Science Foundation could sponsor a research program on social movements to incorporate high levels of participation and partnership by community organizations, and to actively address ethical concerns including research design and data ownership. Funded research of this type could explore forms of knowledge co-production between social movement scholars and social movement actors, providing new understandings of power relationships and knowledge production. This program could also sponsor conferences and workshops for these types of movement-engaged scholars, providing sites for discussion and knowledge sharing among researchers and institutionalizing a network of social movement scholars actively thinking about research ethics. Our model also offers possible contributions that move from social movement scholarship to the field of environmental health sciences, since this reflexive and relational approach is clearly relevant to epidemiologists, public health researchers and exposure scientists who wish to conduct relevant, rigorous work on local health problems.

Acknowledgements

We gratefully acknowledge the participation of our community partners, Alaska Community Action on Toxics, Communities for a Better Environment and the Silent Spring Institute, for their collaboration on numerous research projects and their generous feedback on earlier drafts of this paper. We thank Bindu Panikkar, Leah Greenblum, Tania Jenkins, Mercedes Lyson and Allison Waters for helpful comments on manuscript drafts. We also acknowledge the helpful suggestions of the Editor and two anonymous reviewers. The research described in this paper has been supported by funding from NSF (SES-0822724, SES-9975518, SES-0350691, SES-0450837, SES-0924241), NIH (1R01ES017514), NIEHS (ES017514, ES013258) and the EPA STAR Fellowship (FP-91711901). This paper has not been reviewed by these funding agencies, and the views expressed in this paper are solely the responsibility of the authors. All errors are our own.

Note

1. Exposure research and biomonitoring measure the presence and accumulation of chemicals in household environments, such as dust, or human tissues, such as blood.

References

Adams, C., Brown, P., Morello-Frosch, R., Brody, J. G., Rudel, R. A., Zota, A., Dunagan, S., Tovar, J. & Patton, S. (2011) Disentangling the exposure experience: The roles of community context and report-back of environmental exposure data, *Journal of Health and Social Behavior*, 52(2), pp. 180–196.

Altman, R. G. (2006) Body of evidence: Reproductive health and the environment, Report published by Collaborative on Health and the Environment, Alaska Community Action on Toxics. Available at: http://www.akaction.org/Publications/Health_Care/Body_of_Evidence_Bulletin_v1n1.pdf

Altman, R. G. (2008) Chemical body burden and place-based struggles for environmental health and justice, Ph.D. dissertation, Department of Sociology, Brown University.

Altman, R. G., Morello-Frosch, R., Rudel, J. G., Brown, P. & Averick, M. (2008) Pollution comes home and gets personal: Women's experience of household chemical exposure, *Journal of Health and Social Behavior*, 49(4), pp. 417–435.

Atkinson, P. (2009) Ethics and ethnography, *21st Century Society*, 4(1), pp. 17–30.

Baron, S., Sinclair, R., Payne-Sturges, D., Phelps, J., Zenick, H., Collman, G. & O'Fallon, L. (2009) Partnerships for environmental and occupational justice: Contributions to research, capacity and public health, *American Journal of Public Health*, 3(Suppl. 3), pp. S517–S525.

Bastida, E., Tseng, T., McKeever, C. & Jack, L. (2010) Ethics and community-based participatory research: Perspectives from the field, *Health Promotion Practice*, 11, pp. 16–20.

Beck, U. (1992) *Risk Society: Towards a New Modernity* (London: Sage).

Blee, K. M. & Currier, A. (2011) Ethics beyond the IRB: An introductory essay, *Qualitative Sociology*, 34, pp. 401–413.

Bourdieu, P. (1992) *An Invitation to Reflexive Sociology* (Chicago: University of Chicago Press).

Bourdieu, P. (2000) *Pascalian Meditations* (Cambridge: Polity Press).

Brody, J. G., Morello-Frosch, R., Brown, P., Rudel, R. A., Frye, M., Osimo, C., Perez, C. & Seryak, L. (2007) Improving disclosure and consent: 'Is it safe?', *Journal of Public Health*, 97, pp. 1547–1554.

Brody, J. G., Morello-Frosch, R., Zota, A., Pérez, C. & Rudel, R. (2009) Linking exposure assessment science with policy objectives for environmental justice and breast cancer advocacy: The northern California household exposure study, *American Journal of Public Health*, 99, pp. S600–S609.

Brown, P., Brody, J. G., Morello-Frosch, R., Tovar, J., Zota, A. R. & Rudel, R. A. (2011) Measuring the success of community science: The northern California household exposure study, *Environmental Health Perspectives*. (ahead of print, online December 6, 2011 at http://dx.doi.org/10.1289/ehp.1103734).

Brown, P. & Mikkelsen, E. J. (1990) *No Safe Place: Toxic Waste, Leukemia, and Community Action* (Berkeley: University of California Press).

Brown, P., Morello-Frosch, R., Brody, J., Altman, R. G., Rudel, R., Senier, L., Pérez, C. & Simpson, R. (2010) Institutional review board challenges related to community-based participatory research on human exposure to environmental toxins: A case study, *Environmental Health*, 9, p. 39.

Brugge, D. & Kole, A. (2003) A case study of community-based participatory research ethics: The healthy public housing initiative, *Science and Engineering Ethics*, 9, pp. 485–501.

Bullard, R. (1990) *Dumping in Dixie: Race, Class, and Environmental Quality* (Boulder: Westview Press).

Cable, S. & Cable, C. (1995) *Environmental Problems Grassroots Solutions: The Politics of Grassroots Environmental Conflict* (New York: St. Martin's Press).

Callero, L. P. (2003) The sociology of the self, *Annual Review of Sociology*, 29, pp. 115–133.

Cohen, A., Lopez, A., Malloy, N. & Morello-Frosch, R. (2011) Our environment, our health: A community-based participatory environmental health survey in Richmond, California, *Health Education and Behavior*, online ahead of print July 8, 2011.

Corburn, J. (2005) *Street Science: Community Knowledge and Environmental Health Justice* (Cambridge, MA: MIT Press).

Costello, A. & Zumla, A. (2000) Moving to research partnerships in developing countries, *British Medical Journal*, 321(7264), pp. 827–829.

Dunlap, R. & Catton, W. (1979) Environmental sociology, *Annual Review of Sociology*, 5, pp. 243–273.

Dunlap, R. & Mertig, A. (1992) *American Environmentalism: The U.S. Environmental Movement 1970–1990* (Philadelphia: Taylor & Francis).

Emirbayer, M. (1997) Manifesto for a relational sociology, *The American Journal of Sociology*, 103(2), pp. 281–317.

Emmett, E. & Desai, C. (2010) Community first communication: Reversing information disparities to achieve environmental justice, *Environmental Justice*, 3(3), pp. 79–84.

Epstein, S. (2008) Patient groups and health movements, in: E. Hackett, O. Amsterdamska, M. Lynch & J. Wajcman (Eds) *The Handbook of Science and Technology Studies*, 3rd ed., pp. 499–539 (Cambridge, MA: MIT Press).

Flicker, S., Travers, R., Guta, A., McDonald, S. & Meagher, A. (2007) Ethical dilemmas in community-based participatory research: Recommendations for institutional review boards, *Journal of Urban Health*, 84(4), pp. 478–493.

Gerhardstein, B. & Brown, P. (2005) The benefits of community medical monitoring at nuclear weapons production sites: Lessons from Fernald, *Environmental Law Reporter*, 35, pp. 10530–10538.

Giddens, A. (1984) *Constitution of Society: Outline of the Theory of Structuration* (Berkeley, CA: University of California Press).

Giddens, A. (1991) *Modernity and Self-Identity: Self and Society in the Late Modern Age* (Cambridge: Polity Press).

Giese-Davis, J. (2008) Community/research collaborations: Ethics and funding, *Clinical Psychology: Science and Practice*, 15(2), pp. 149–152.

Gottlieb, R. (1993) *Forcing the Spring: The Transformation of the American Environmental Movement* (Washington, DC: Island Press).

Hammersley, M. & Atkinson, P. (2007) *Ethnography: Principles in Practice* (London: Routledge).

Israel, B. A., Schulz, A. J., Parker, E. A. & Becker, A. B. (1998) Review of community-based research: Assessing partnership approaches to improve public health, *Annual Review of Public Health*, 19, pp. 173–202.

Kenway, J., & McLeod, J. (2004) Bourdieu's reflexive sociology and 'spaces of points of view': whose reflexivity, which perspective? *British Journal of Sociology of Education*, 25(4), pp. 525–544.

Mayer, B. (2008) *Blue-Green Coalitions: Fighting for Safe Workplaces and Health Communities* (Ithaca, NY: Cornell University Press).

Minkler, M. (2004) Ethical challenges for the 'outside' researcher in community-based participatory research, *Health Education & Behavior*, 31, pp. 684–697.

Minkler, M. & Wallerstein, N. (2008) *Community-Based Participatory Research for Health: From Process to Outcome* (San Francisco: Jossey-Bass).

Mohai, P. & Bryant, B. (1992) Environmental injustice: Weighing race and class as factors in the distribution of environmental hazards, *University of Colorado Law Review*, 63, p. 921.

Mohai, P., Pellow, D. & Roberts, J. T. (2009) Environmental justice, *Annual Review of Environment and Resources*, 34, pp. 405–430.

Morello-Frosch, R., Brody, J., Brown, P., Altman, R., Rudel, R. & Pérez, C. (2009) Toxic ignorance and right-to-know in biomonitoring results communication: A survey of scientists and study participants, *Environmental Health*, 8(1), p. 6.

Morello-Frosch, R., Brown, P., Brody, J. G., Altman, R. G., Rudel, R. A., Zota, A. R. & Perez, C. (2011) Exerts, ethics, and environmental justice: Communicating and contesting results from personal exposure science, in: G. Ottinger & B. Cohen (Eds) *Environmental Justice and the Transformation of Science and Engineering*, pp. 93–118 (Cambridge, MA: MIT Press).

Morello-Frosch, R., Pastor, M., Sadd, J., Porras, C. & Prichard, M. (2005) Citizens, science, and data judo: Leveraging community-based participatory research to build a regional collaborative for environmental justice in southern California, in: B. A. Israel, E. Eng, A. J. Schulz & E. A. Parker (Eds) *Methods for Conducting Community-Based Participatory Research in Public Health*, pp. 371–391 (San Francisco: Jossey-Bass).

Morello-Frosch, R., Zavestoski, S., Brown, P., Altman, R. G., McCormick, S. & Mayer, B. (2006) Embodied health movements: Responses to a 'scientized' world, in: S. Frickel & K. Moore (Eds) *The New Political Sociology of Science: Institutions, Networks, and Power*, pp. 244–271 (Madison: University of Wisconsin Press, Madison).

Murphy, E. & Dingwall, R. (2001) The ethics of ethnography, in: P. Atkinson (Ed.) *Handbook of Ethnography*, pp. 339–351 (London: Sage).

O'Fallon, L. & Dearry, A. (2002) Community-based participatory research as a tool to advance environmental health sciences, *Environmental Health Perspectives*, 110(S2), pp. 155–159.

Pillow, W. (2003) Confession, catharsis, or cure? Rethinking the use of reflexivity as methodological power in qualitative research, *International Journal of Qualitative Studies in Education*, 16(2), pp. 175–196.

Schirato, T. & Webb, J. (2003) Bourdieu's concept of reflexivity as metaliteracy, *Cultural Studies*, 17(3/4), pp. 539–552.

Senier, L., Hudson, B., Fort, S., Hoover, L., Tillson, R. & Brown, P. (2008) Brown superfund basic research program: A multistakeholder partnership addresses real-world problems in contaminated communities, *Environmental Science & Technology*, 42(13), pp. 4655–4662.

Smith-Doerr, L. (2006) Learning to reflect or deflect? U.S. policies and graduate programs' ethics trainings for life scientists, in: S. Frickel & K. Moore (Eds) *New Political Sociology of Science*, pp. 405–431 (Madison: University of Wisconsin Press).

Swauger, M. (2011) Afterword: The ethics of risk, power, and representation, *Qualitative Sociology*, 34, pp. 497–502.

Watkins, B. X., Shepard, P. & Corbin-Mark, C. (2009) Completing the circle: A model for effective community review of environmental health research, *American Journal of Public Health*, 99, pp. S567–S577.

Wu, N., McClean, M., Brown, P., Aschengrau, A. & Webster, T. (2009) Participant experiences in a breastmilk biomonitoring study: Qualitative assessment, *Environmental Health*, 8, p. 4.

Ethical and Political Challenges of Participatory Action Research in the Academy: Reflections on Social Movements and Knowledge Production in South Africa

MARCELLE C. DAWSON & LUKE SINWELL

Sociology Department and South African Research Chair in Social Change, University of Johannesburg, Johannesburg, South Africa

ABSTRACT   From the vantage point of South Africa, this article highlights a number of ethical challenges that could potentially arise in the relationship between social movement researchers and activists in the pursuit of social justice and transformation. In contrast to conventional approaches to social science more generally, we argue that a neat separation between theory and action is useful neither for producing knowledge within the academy nor for advancing the causes of social movements. The article reflects on two different research experiences in order to explain the limitations and quandaries that confront academics who seek to negotiate scholar–activist identities. In doing so, it extends the work of Croteau, who has explored the tensions between activism and scholarship. Drawing on participatory action research (PAR) approaches, including Touraine's method of sociological intervention, we suggest that a refined approach to PAR may assist in countering the inequalities that have been created in the academy between the researcher and researched, thereby alleviating some of the ethical and political concerns that inevitably confront scholar-activists.

Introduction

Social movement researchers who employ research techniques aimed at exposing social inequalities and who seek to actively promote progressive social change are often confronted with the task of having to bridge the gap between two arenas that seem to be at odds. On one hand, the academy can be dismissive, even contemptuous, of their 'activist research' pursuits. On the other, they may be accused of having very little to offer activists by way of real grassroots struggle. It comes as no surprise that the very nature of the academy, with all its neoliberal conventions and its relationship with big corporations (Aronowitz, 2000; Callinicos, 2006; Pithouse, 2006), has made it very difficult for scholars to succeed in their pursuit of mutually rewarding and productive relationships with social movements that aim to change the social order so that the majority of the population—as opposed to an elite—can thrive.

Rethinking the practical value of academic work involves a re-articulation of one's identity. Self-identification significantly shapes the rationale, process and outcome of any research investigation and will influence how one understands who and what knowledge is for. Drawing on the findings of her study of the conflict between activist research and academic success, Cancian (1993, p. 92) claimed that scholars with a primarily *academic* research focus displayed an 'emotional detachment' from the people being studied, while they enjoyed 'close ties with faculty and students'. Croteau (2005, p. 32) used the label 'SCHOLAR–activist' to refer to such academics who grant primacy to their careers. These types of scholars tend to produce work '*about* social movements' but very rarely is their work 'useful for movements' because it tends to 'involve broad social criticism or may focus narrowly on social problems without considering how social change might occur' (Croteau, 2005, p. 33, emphasis in original).

The 'scholar–ACTIVIST' approach, whereby academics 'prioritize activism at the expense of scholarship' (Croteau, 2005, p. 35), represents an alternative path. However, since this research strategy may lead to a situation in which scholars are judged harshly by the academy for failing to advance the 'intellectual frontiers' of their academic disciplines and for exercising an inappropriate balance between scholarship and activism (see Cancian, 1993, p. 102), it is unlikely to be pursued by emerging scholars whose careers are at stake. Even established academics who already have permanent positions at universities, and thus have less to lose, may not want to tarnish their reputations or reduce their chances of being awarded lucrative research grants.

The awarding of publication subsidies is a particular feature of the South African higher education system that discourages the pursuit of the scholar–ACTIVIST approach: South African universities receive state-sponsored subsidies for 'accredited' research outputs. In particular, academics are being 'incentivised' to produce journal articles (as opposed to book chapters) and are pressured into submitting their work only to those journals that appear on the lists of accredited journals, which are sent annually to the universities. At most of the universities, a portion of this money filters down to the individual authors who may then make claims on the funds that their work has generated, provided the money is being used for research-related expenses.[1] This research accreditation system has been harshly criticised by scholars who feel that the 'motive of writing articles for money is wrong, wrong, wrong! It's morally wrong and it is not likely to produce anything of any standard' (Gray cited in Byrne, n.d.). Others have suggested that the system 'serves to commoditize scholarly endeavour and foster slothful, meretricious, even cynical intellectual habits. Academics feel constantly under pressure to rush into print work which is inadequate, incomplete or superficial' (Cornwall cited in Byrne, n.d.). Moreover, as Mbali (2011) points out, '[e]mployees who do not produce any [accredited] articles get threatened with "performativity" interviews, loss of sabbaticals, no promotion and early retirement' (see also Mbali, 2010). Although not set in stone, publication targets may be specified. For example, scholars in teaching jobs may be expected to produce one accredited publication per annum, while those based at research centres may be required to publish three or four 'units' per year. Faced with these constraints, it is virtually impossible for South African scholars to pursue a scholar–ACTIVIST approach without coming under scrutiny by university bureaucrats for whom research productivity is measured largely in terms of how much money the individual scholar earns for the institution.

There is a third possible path for those who want to make research efforts 'accountable to both activist and academic standards' (Cancian, 1993, p. 93). The 'two career strategy' (Cancian, 1993, p. 103) involves scholars having 'a second career oriented to social change on top of their academic publications and activities'. For Croteau (2005, p. 35), this type of strategy—what he calls the SCHOLAR–ACTIVIST trajectory—entails the ability (and no doubt the desire) 'to produce two versions of [one's] scholarship—one for consumption by colleagues and a translation for a more activist audience'. However, besides the additional demands on the time and energy of those who adopt this strategy (Cancian, 1993, p. 103), there is the more insidious problem of a *parallel*, rather than integrated involvement in the worlds of the academy and activism. It is as though the individual is required to lead a double life, in which the engagements in the one arena are divorced from the activities in the other, and one may fall into the trap of having to 'divide [one's intellectual work] into two parts' or produce 'one kind of work for the "simple" in ... movements and one for the elite in the universities' (Cox & Fominaya, 2009, p. 10 citing Gramsci). This type of approach, we argue, exacerbates the divide between academics and activists. Within the academy it treats the institutional criteria for measuring the acceptability and validity of research outputs as a given and, amidst the activist world, assumes that academic knowledge must be dumbed down by academics for consumption by movement activists. Activist knowledge is merely considered 'data' until it is analysed, codified and written-up by the academic. In other words, activists' knowledge is rarely recognised as an important form of knowledge in its own right.

The theoretical and typological frameworks offered above provide a useful context in which to situate this debate on ethics and methods in social movement research in South Africa. They help to explain the orientation of researchers to their projects and to the participants who are involved in their investigations. Against this background, we illustrate the pitfalls of erring too heavily on the side of either the scholar or the activist. The cases discussed below stem from our personal research experiences. They highlight a number of ethical considerations that have encouraged us to approach social movement research in different ways and to rethink the process of knowledge production within the academy. Over the short course of our academic careers we have straddled the divide between academia and activism. We have shifted our approaches over time, and this has shaped the way in which we identify with the academy, on the one hand, and as activists on the other. The primary ethical consideration that has driven our research is that it should contribute to progressive social change and the advancement of social movements.

A range of novel methodological strategies have been employed by researchers in order to overcome the ethical dilemmas that they face. In this paper, we offer a critical reflection of two of these, namely participatory action research (PAR) and sociological intervention as proposed by social movement researcher, Alain Touraine. Several authors, including Touraine, have created a false binary between knowledge production in the academy and activism in social movements which, as suggested above, obscures the fact that knowledge is also produced in social movements.

Based on our research experiences, we argue that existing attempts to merge scholar and activist identities are limited and we suggest that to be simultaneously effective in both arenas, scholarship should not to be separated from activism. In order to contextualise the discussion of our own experiences, it is necessary to provide some insights into the national context, with particular reference to the emergence of, and scholarship on, social movements in contemporary South Africa.

Movements, Intellectuals and Scholarship in Post-Apartheid South Africa

In the mid-1990s, following the accession to power by the African National Congress (ANC), there was a brief respite as far as popular movements and community resistance were concerned. People had faith that the ANC would deliver on the promises contained in the pro-poor Reconstruction and Development Programme (RDP). However, the apparent hiatus in grassroots struggle did not last very long. Shortly after its rise to power in 1994, the ruling ANC began to adopt a range of neoliberal policies, epitomised by the Growth, Employment and Redistribution Programme (GEAR), which replaced the RDP as an economic strategy. As a consequence, poverty and unemployment continue to mar South Africa's young democracy. A number of scholars have referred to democratic change as an 'elite transition' that has benefitted a minority and entrenched inequality (Bond, 2000a, 2000b; Seekings & Nattrass, 2006). Others have characterised South Africa as a 'low intensity' or 'fragile' democracy, suggesting that class and race fault lines have prevailed in the post-apartheid period and that they continue to threaten the country's stability (Marais, 2001, 2011; McKinley, 2003, 2006a, 2006b; Beall et al., 2005; Dwyer, 2009). Indeed, with a Gini-coefficient of 0.679[2], South Africa—now ahead of Brazil—is one of the most unequal countries in the world (Davies, 2011).

The late 1990s saw the emergence of a range of social movements that were protesting against the dire socio-economic conditions still facing the majority of the population. Economic redistribution was a key claim around which several of these movements framed their demands. Importantly, the Congress of South African Trade Unions (COSATU) and the South African Communist Party (SACP), despite being part of the ruling alliance, were openly critical of GEAR. However, more contentious and sustained opposition has emerged from the social movements outside of mainstream politics. Some of the more resilient movements include the Anti-Privatisation Forum (comprising about 30 affiliates), Western Cape Anti-Eviction Campaign, Landless People's Movement (LPM; although it went through a significant decline from 2004 to 2009), Abahlali base Mjondolo (a shack dwellers organisation), Jubilee and the Treatment Action Campaign[3]. The support base of these movements is overwhelmingly black and poor, and many of the community activists are unemployed. There is, however, a handful of white and middle class supporters.

Existing literature on South African social movements tends to be largely descriptive and is based on individual case studies of movements or of movement issues, for example, access to land, basic services and healthcare (see Ballard et al., 2006 and Robins, 2008). Other writings have produced somewhat romanticised accounts of these movements (see Gibson, 2006). Activist-intellectual, Ashwin Desai (2006, p.8) noted that these texts rely on evoking pity for the poor, rather than interrogating the relationship between the researcher and the actor or between academic research findings and possibilities for practical social change. These tensions were reflected in South African social movement discourse. The early period of writing about movements was done largely by university-based academics and independent scholar–ACTIVISTS. Both sets of individuals, many of whom are white, were directly involved in shaping the character of the movements as far as ideology and tactics were concerned. It was not long before this group of authors became the target for vitriolic attacks from within and outside the academy, but especially by activists who criticised some of the movements for having a colonial mentality and who accused the white, educated leaders of trying to dominate and control the poor black followers in their pursuit of academic interests. Much of this critique was exaggerated and some of it was unfounded, but it had an impact on the practice of social movement scholarship in South Africa. Fearing a backlash from the movements, some scholars began to retreat into the SCHOLAR–activist mode, while others depicted the poor as virtuous and portrayed their voices as the embodiment of the truth (see Desai, 2006). Still others took the criticisms head on and began to engage in a scholarly fashion with ethics and the relationship between researchers and activists. In particular, the work by Desai (2006) and Walsh (2008) has made a valuable contribution to articulating a SCHOLAR–ACTIVIST identity within a South African context.

Desai (2006, p. 5) likened elitist knowledge production that is in no way beneficial to social movements to 'spying'. He (2006, p. 5) asked: 'Is it not patronizing to presume to label the politics of those we consort with in struggle in academic texts but not to engage them in an exchange of ideas over these same issues?' Desai (2006, p. 9) invited movement activists to enter academic spaces and to engage in their own analyses of how we (scholars) have analysed their movements.

Reflecting on Desai's work and also on her own position as a white foreigner who was actively involved in the struggles of a community of black shack dwellers in South Africa, Walsh (2008) contested constructions of the 'poor' by academics, and also challenged essentialist identities that often set the terms for mobilisation and introduced the term 'uncomfortable collaborations' to explore possibilities for engagement between heterogeneous groups that 'can burst open geographic and identity-based alliances, deterritorialising groupings around commonalities of desire, struggle and event' (Walsh, 2008, p. 256). For Walsh, '[t]hese collaborations are … sites of friction in which diverse power struggles and contestation at the local, everyday level arise'.

During an informal discussion on the issue of the relationship between scholars and activists, Dale McKinley, seasoned activist-intellectual and independent writer in South Africa, suggested that academics tend to be 'reactive' rather than 'proactive' in the sense that they are quick to rush to sites of activism after the fact and to pontificate about the perceived causes of social protest or elaborate on the experiences of demonstrators (McKinley, Personal communication, 11 January 2011). He went on to criticise conventional research methods, or what he refers to as 'parachute research', where researchers dip into movements to get little bits of information and then shoot off again into their ivory towers to construct elaborate arguments or theories that are disconnected from the reality of the movements' politics and mobilisation efforts. McKinley lamented the lack of interest in, and sustained commitment to South Africa's movements by senior academics in particular (McKinley, Personal communication, 11 January 2011). He suggested that these academics usually present themselves at the scene of high-profile demonstrations that receive a great deal of media attention, often as a means of self-promotion as 'public intellectuals'. However, when the heat dies down, the academics retreat into their offices until the next big event. In order to address some of the pitfalls of our own scholarship, and to expand on the frameworks discussed above, we elaborate on two personal research experiences that have encouraged us to map a way forward for a tighter, more workable integration of scholar and activist identities.

Ethical Challenges in Social Movement Research: Lessons from South Africa

Marcelle C. Dawson

During an investigation into the policing of protest in South Africa (Dawson, 2010), the potential advantages of a more participatory approach were thrown into stark relief. I was approached by a Dutch NGO to research the issue of police repression in relation to social movement activists in South Africa. I had about six months to complete the investigation and, mindful of this time limitation, I felt that a conventional qualitative approach to research, in which I led the process from start to finish, would be the most efficient method. One of the fieldwork trips was to a rural village in Limpopo province that had been severely affected by mining exploits in the area. During demonstrations against the mining companies, several activists were badly beaten by police and some were arrested. In one of the very first focus groups conducted in this area, a participant urged, 'It will please me if you could remember that you have heard our grievances and perhaps come up with solutions to our problems' (Focus group, 26 September 2009, translated from Pedi). Shortly afterwards, an elderly gentleman stood up and said,

> We are giving information to you but we do not know the way forward. We do not know if we are going to get help or not. We've had too many people coming to our village asking what you are asking. We tell them what they need to know but there is no way forward. Even now, we are giving the information but you are going to file it. You hurt us by doing so. (Focus group, 26 September 2009, translated from Pedi)

It struck me that I had a very different research agenda to that of the activists. They expected me to do something practical with the information I was gathering, while my motivation was to gather and analyse data in order to draw theoretical conclusions about the policing of protest. Indeed, I was lambasted by an activist for using my scholarship only to advance my own career. In an interview with an activist in the Western Cape, I was asked, 'What do you want to know? You researchers always come here asking questions but when we ask questions you give us theories. That's what you are doing. You are theorising the shit out of people's pain!' (Poni, Interview, 10 September 2009). So, the first set of ethical dilemmas related to the issue of what we do with our research.

The second ethical issue I faced during this investigation was an attempt by one of the activists in Limpopo to 'hijack' my research agenda. He acted as a gatekeeper and, in-between interviews, grilled me on the kinds of questions I was asking about police brutality and insisted instead that I ask different sorts of questions about the problems faced by the communities whose rights were being violated by the mining companies. While this is a valid and necessary endeavour, it was not the focus of my research. The fieldwork trip ended on quite a sour note, with the activist in question reprimanding me for not taking his suggestions on board. I found the experience rather unsettling. This aspect of my research illustrates an ethical concern that has to do with who we involve in the research process and how we go about doing this. Later in the discussion we reflect on these two ethical questions as well as the concerns raised in the second example, showing how we can possibly adapt research methods in such a way so as to resolve the ethical quandaries faced by social movement researchers.

Luke Sinwell

In my Ph.D. and other academic work in poor communities in South Africa, I have traced the historical process through community-based organisations mobilised collectively to demand access to water, electricity and housing. The theory that I produced was significantly grounded in movements and local knowledge since it was based on the realities of activists. However, given the style of writing required for peer-review publications, it was not directly applicable to activists. The shortcomings of the SCHOLAR–activist approach were evident in my work and I wanted to find a way to bring out the ACTIVIST dimension more strongly. Seeking to connect the theory to actual practices, I initiated what can be called an activist forum. A colleague and I brought 10 activists from various protesting communities that we were working with to the university on several occasions and began planning for a large coordinated march that would unite the various organisations under a common development banner.

However, reality soon put a damper on our plans as affiliates of the LPM, who we were working with closely at the time, began to experience severe state repression. For example, in the community of Etwatwa, east of Johannesburg, residents held a mass meeting demanding that their local government councillor step down. The councillor called the police who then proceeded to shoot three residents with live ammunition, injuring two and killing one. This was followed by the arrest of several residents. Around the same time, another battle over electricity between middle-class residents and shack dwellers broke out in Protea South (another LPM affiliate in Soweto) and two people were shot.

These brutal events dominated the discussion when we met again to discuss the way forward. In the end, we decided not to press for the united march because we did not know how we would respond to possible attempts by the police to clamp down on our actions. Our ability and desire to coordinate mass action, without necessarily being directly affected by the consequences of that action, raised important ethical questions. We were forced to consider Juris's point that activist-intellectuals must be willing to 'put their body on the line during direct action' (Juris, 2007, p. 165). At the time, I was convinced that the researcher may need to 'live the emotions associated with direct action' (*ibid.*), or otherwise risk distancing themselves from the strategic actions that they have planned for. We must not expect poor people to fight battles for us while we decide when to be the observer and when we will be the *participant* observer.

This experience epitomised the scholar–ACTIVIST approach, which paid little attention to theorisation and the production of knowledge, and thereby created a sharp distinction between reflection (by researchers) and action (by activists). But, this proved insufficient. The quest instead should involve eroding the hierarchy between movements and the academy. This means recognising that thought and action are intertwined and therefore that activism and knowledge production do not necessarily occur separately. An ethical approach to 'sociological intervention' involves acknowledging that movements themselves are the sites of the production of theoretical and sociological knowledge. Indeed, actors within movements make decisions over a period of time, reflecting (not merely acting) and deciding upon the most appropriate action.

With these experiences and their concomitant ethical conundrums in mind, the discussion now considers PAR and Touraine's method of sociological intervention, paying attention to the ways in which these methods can potentially resolve the kinds of ethical problems we experienced, but also highlighting the shortcomings of these approaches. While these methodological 'solutions' may help to resolve the epistemological problems associated with researching social movements, there are also some practical solutions that may help to alter the structural environment in which academics operate in order to facilitate the development of a more integrated scholar–activist approach to research.

Resolving Epistemological and Structural Constraints: Methodological and Policy Adaptations

PAR and Sociological Intervention

Efforts to develop research methodologies that could potentially reduce tensions between researchers and activists and equip researchers with the tools to bring about progressive social change have a long history. The backlash against traditional approaches to research that advocate the separation of knowledge from practical efforts at achieving social change gained prominence in the 1960s and 1970s when radical adult educators in Asia, Latin America and Africa embarked on the method of PAR (Kindon *et al.*, 2007). Their aim was to use the expertise, theory and research of 'outsiders' to empower the oppressed and marginalised, particularly in the third world (Fals-Borda & Rahman, 1991). PAR involves engaging in a collaborative process of reflection, idea generation, action and writing with research participants throughout the research process.

Paulo Freire, most well known for his literacy campaigns with poor peasants in Brazil, was a pioneer of PAR who sought to empower marginalised people to take greater control over their own lives. This meant that people needed to be 'conscientised' so that they could overcome the structures of society that served to oppress them (Freire, 1972). Following Gramsci, PAR proponents such as Freire aimed to transform: '"common" sense into "good" sense or critical knowledge that would be the sum of experiential and theoretical knowledge' (Fals-Borda & Rahman, 1991, p. 9). This approach was drawn from activists' concerns with the colonisation of knowledge, and it sought to offset imperialism and capitalist expansion in the third world. PAR represented the potential for research to be used to create radical changes in society.

Similarly, social movement studies emerged in the 1960s, at the peak of the civil rights and anti-war movement in the USA, in part to validate the knowledge and practices of activists struggling for justice. In response to the perception that activists engaging in extra-institutional politics had a mob mentality, this era of social movement studies viewed movements as 'collective efforts to pursue interests with intelligible strategies and rational goals' (Flacks, 2005, p. 6). Critically, Flacks indicates that out of this period emerged a consensus that:

> If your research was focused on the relatively powerless and disadvantaged, you had an ethical obligation to enable them to use the results; [t]hat one ought to be sensitive to the possible ways your work could be used to perpetuate established social arrangements and repress opposition; [t]hat the study of social movements ought to provide movement activists with intellectual resources they might not readily obtain otherwise (Flacks, 2005, pp. 6–7).

It should, however, be noted that with the institutional mainstreaming of bottom-up approaches to development in the 1990s, PAR was watered-down into something called participatory research or participatory poverty assessments, which aimed to help organisations such as the World Bank to undertake development projects that were more locally appropriate. Rather than 'conscientizing' the marginalised to change their structural conditions in the world, research was done mostly 'on' (rather than with) the marginalised 'to provide policy-makers with information about poor people's perspectives on poverty' (Brock, 2002, p. 1). Moreover, as Croteau (2005, p. 34) points out PAR 'emerged from active civil society and social movement bases' and was inserted later into the academy. Considering the limitations and prescriptions of the academy, PAR runs the risk of 'los[ing] much of its militant edge' and 'diverg[ing] significantly from more direct challenges to power posed by social movement efforts'. Thus, while the research process may be more inclusive, PAR does not guarantee that research outcomes will be useful for grassroots struggles.

Despite its limitations, PAR remains an influential paradigm in social movement research. The work of Alain Touraine (1980, 1981) relied to a certain extent on some of the principles of PAR and his contributions are particularly valuable to a discussion on ethics in social movement research. Touraine (1980, 1981) regarded the actions of social movements as critical in the process of creating and changing social systems. He was absorbed by the question of *how* to study social movement actors and, to this end, he developed the method of 'sociological intervention' (Touraine, 1980, p. 8). In the vein of PAR, Touraine urged that 'the actor must take part in the research *as an actor* and not as a subject for observation or experimentation' (Touraine, 1980, p. 10, emphasis added). Sociological intervention was conceived of as a series of lengthy interactions between actors and researchers in which '[t]he object of the analysis should not be the behaviour of the actor, but the analysis which the actor makes of his [or her] own behaviour and of the behaviour of his [or her] social partners' (Touraine, 1980, p. 11). Following the process in which the actor becomes a self-analyst (Touraine, 1980, p. 10), the researcher stages an intervention by introducing a series of interlocutors who hold opposing points of view to the actors. The idea behind the involvement of these interlocutors is to encourage the actors to step back from their own actions and ideology and see how their actions in concert with the actions of others can potentially lead to social change (Hamel, 1998, p. 3; Brincker & Gundelach, 2005, p. 366). In contrast to PAR methods aimed at bringing about change, the purpose of the intervention method 'is to create an understanding among collective actors of their potential for and role in social change' (Brincker & Gundelach, 2005, p. 369). Following the intervention, the researcher reveals to the actors his or her interpretation of the self-analysis put forward by the actors. The actors then have a chance to engage with the researcher and to interrogate and challenge his or her interpretations and to use these interpretations to inform their subsequent actions. These ongoing, mutually influential exchanges between analysis and action constitute what Touraine (1980, 1981) referred to as 'permanent sociology'.

Locating the intervention method among other types of research techniques, Touraine (1980, p. 12) was adamant that:

> [t]he sociologist in no way identifies himself with the actual struggle of the actor or with his ideology. Nor is he a neutral observer incapable of interacting with the actor without destroying him. He acts as an agent in the analysis of the actor and his

intervention enables him to advance his own analysis. This role of the researcher is very far from the cold objectivity of the sociological tradition. But it is even further from the identification with the actor, which the different types of militant research or action-research propose. The researcher aims at knowledge while the actor's aim is action.

There are a number of limitations to this approach: In practice, there may be time and resource constraints that hamper a neat replication of the sociological intervention method as Touraine intended it. Second, the nature of a social movement and the relationships between its activists and the movement's opponents may preclude a researcher from staging an intervention such as that described by Touraine. Third, there is no guarantee that activists will actually translate the researchers' analysis into action, no matter how collaborative the process was. Before reflecting on some of the ways in which these methodological apparatuses can be adapted so as to mitigate the ethical difficulties addressed above, we consider, very briefly, some practical suggestions that may eventually alter the context and conventions that inform knowledge production in the academy.

Altering Academic Conventions

As social movement researchers seek to reduce the gap between activists and academics, we contend that it is not enough merely to amend our methodological approaches. We should also begin to push the boundaries of the academy from within. In other words, social movement researchers should use the academy as a site of struggle to challenge academic prescriptions that make the academy inaccessible to those on the outside. Changes to our research methods alone will lead to the limited versions of SCHOLAR–ACTIVIST explored above, where researchers feel compelled to adopt the dual career approach. Without concomitant changes to the way in which the academy operates, researchers will not be in a position to sufficiently integrate their scholarship with their activism and will continue to be pulled in two separate and demanding directions.

An adequately integrated SCHOLAR–ACTIVIST approach entails redefining the academy's expectations of social movement research in such a way that the research 'outputs' are actually relevant and useful for activists, but are also regarded as a valid contribution to scholarly work. A major obstacle in this regard is that the kind of language deemed acceptable within the academy is often indecipherable to movement activists owing, in large part, to the use of disciplinary jargon and theoretical terminology that could be explained in far simpler ways. We are not suggesting that 'anything goes' but, as things currently stand, activists are considered to be an important *source* of information, yet they are not given the same recognition as *evaluators* of the knowledge that is produced from information that they themselves have generated. One possible way of addressing this problem is to insist on a peer-review system in which one of the referees is a movement-based—as opposed to university-based—activist. After all, activists are quite capable of evaluating the quality and accuracy of knowledge about social movements. This may involve scholars giving up some of the power and privilege that they enjoy in the academy, but it would also be a step towards advancing social justice. Second, and related to the first point, collaborative writing endeavours between academics and activists should be encouraged. Linking this suggestion to the third point, scholars in South African higher education institutions will become less driven by the pursuit of solely authored publications if the system of receiving government subsidies for accredited publications were scrapped.

Although speculative, these suggestions are grounded in the experiences of doing social movement research in the South African context and we contend that a shift in the prescriptions that currently govern knowledge production in the academy will alter the context in which social movements do research, thereby enabling a more valuable relationship between the scholar and activist aspects of their identities.

Mapping a Way Forward for the SCHOLAR–ACTIVIST Approach

Based on the lessons learned from our research experiences, and in view of the shortcomings of PAR and the method of sociological intervention, we bring the discussion to a close with a few brief reflections on how to advance the SCHOLAR–ACTIVIST approach. First, while the SCHOLAR–activist approach risks producing abstract theory that may be useless to the movements under investigation, the researcher should also not commit the mistake of fuelling action at the expense of critical reflection on and within movements. Indeed, the success of movements does not depend only on their ability in the battlefield. If we choose the right research questions and connect adequately to the movements we investigate, then putting our heads to good use may prove more useful than putting our bodies on the line. Thus, the approach that we take is one in which the researcher can be involved with the movements they are studying but at enough of a distance so as to avoid the pitfall of being so biased that we become blind to the problems within the movements that we identify with. Rather, involvement in, and even sympathy for, a movement can enable one to come to grips with the internal contradictions within a movement or the structural limitations with which they are faced. Researchers and movements are more likely to be able to constructively engage each other if they are on the same side. Being involved in the everyday activities of a movement can enable academics to undertake research projects that are valuable to activists.

Second, an ethical approach to the study of social movements, in our view, strives for a situation in which the researcher does not have to live out—in separate spaces—the desire to be a successful scholar who is, at the same time, involved in progressive social change. While we acknowledge the value of certain aspects of Touraine's method, we diverge from his suggestion that the researcher must use the analysis produced by activists in theory building within the academy and that activists should use the researchers' analysis in their action. We argue that in order to shift the balance of power in the relationship between scholars and activists, researchers should use the academy as a site of social change so that activists' knowledge is recognised as a valid form of theorising and idea generation.

Third, an integrated SCHOLAR–ACTIVIST approach guards against a situation in which academics become stereotyped as 'knowledge-producers', while those outside the academy are seen as 'knowledge-users'. As Barker and Cox (2002, p. 1) astutely remind us, 'movements and their activists also produce "knowledge"'; they are not merely sites of activism. Sharing these sentiments, and drawing on their study of the anti-genetic engineering movement, Schurman and Munro (2006) showed how the intellectual work of a small group of activists advanced the movement and its cause. They suggested that the activists 'forged an oppositional ideology and concrete set of grievances upon which a movement could later be built' (Schurman & Munro, 2006, p. 4).

For Barker and Cox (2002, p. 2), narrow academic theorising runs the risk of 'treat[ing] what are, precisely, *movements* as static "fields"' (Barker & Cox, 2002, p. 2, emphasis in original). The fourth point is, thus, that an amalgamated SCHOLAR–ACTIVIST approach involves ongoing 'fieldwork'; research in motion, involving continuous conversation—along the lines of Touraine's permanent sociology—between scholars and activists with the aim of being *collectively* involved in social change. Research conducted in this way ensures that there is a mutually influential relationship between theory and action. We, therefore, agree with the assertion that:

> Activism uninformed by broader theories of power and social change is more likely to fall prey to common pitfalls and less likely to maximise the potential for change. Social movements without access to routine reflection on practice are prisoners of their present conditions. Theory uninformed by and isolated from social movement struggles is more likely to be sterile and less likely to capture the vibrant heart and subtle nuances of movement efforts. Theorists without significant connections to social movements can end up constructing elegant abstractions with little real insight or utility (Croteau *et al.*, 2005, p. xiii)

Having argued that a SCHOLAR–ACTIVIST approach should go further than attempting to 'succeed' in both arenas by taking on a double workload, we conclude—borrowing from Juris (2007)—that an integrated SCHOLAR–ACTIVIST approach to social movement research entails 'militant ethnography', which Juris (2007, p. 164) sees as 'a politically engaged and collaborative form of participant observation carried out from within rather than outside grassroots movements' This role, or what some authors have referred to as the 'activist-intellectual', requires scholars to be embedded and active within movements *and* to challenge the 'institutional logic of academia itself' (Juris, 2007, p. 165, emphasis added. See also Juris, 2008, p. 23–24). This kind of methodological approach aims to resolve the ethical problems that we raised in this discussion by subverting the power dynamics between activists and academics and promoting activist theorising in the pursuit of social justice.

Acknowledgement

The authors would like to thank Kevin Gillan for his valuable comments on this paper.

Notes

1. For more information on the subsidy formulas employed by the University of Johannesburg, where the authors are based, refer to UJ, n.d. 'DHET publication subsidy' which can be accessed at http://www.uj.ac.za/EN/Research/Research%20Information/IncentivesforResearchers/Pages/DoEPublicationSubsidy.aspx
2. The Gini-coefficient is a statistical measure of wealth or income inequality. It can range from 0 to 1, where 0 is indicative of an equal distribution of wealth, while a higher Gini-coefficient indicates a higher level of inequality.
3. Some scholars contest the idea that the TAC is a grassroots social movement and argue that it operates as a well-funded NGO or pressure group interested in lobbying the government rather than mobilising for radical and progressive social change.

References

Aronowitz, S. (2000) *The Knowledge Factory: Dismantling the Corporate University and Creating True Higher Learning* (Boston: Beacon Press).

Ballard, R., Habib, A. & Valodia, I. (Eds) (2006) *Voices of Protest: Social Movements in Post-Apartheid South Africa* (Durban: University of KwaZulu-Natal Press).

Barker, C. & Cox, L. (2002) What have the Romans ever done for us? Academic and activist forms of movement theorizing. In C. Barker and M. Tyldesley (Eds) *Alternative futures and popular protest VIII: Conference proceedings*, Manchester: Manchester Metropolitan University.

Beall, J., Gelb, S. & Hassim, S. (2005) Fragile stability: State and society in democratic South Africa, *Journal of Southern African Studies*, 31(4), pp. 681–700.

Bond, P. (2000a) *Cities of Gold, Townships of Coal: Essays on South Africa's New Urban Crisis* (Trenton, NJ/Asmara, Eritrea: Africa World Press).

Bond, P. (2000b) *Elite Transition: Globalisation and the Rise of Economic Fundamentalism in South Africa* (London/Pietermaritzburg: Pluto/University of Natal Press).

Brincker, B. & Gundelach, P. (2005) Sociologists in action: A critical exploration of the intervention method, *Acta Sociologica*, 48(4), pp. 365–375.

Brock, K. (2002) Introduction: Knowing poverty: Critical reflections on participatory research and policy, in: K. Brock & R. McGee (Eds) *Knowing Poverty: Critical Reflection on Participatory Research and Policy*, pp. 1–13 (London and Sterling: Earthscan Publishers).

Callinicos, A. (2006) *Universities in a Neoliberal World* (London: Bookmarks).

Cancian, F. (1993) Conflicts between activist research and academic success: Participatory research an alternative strategies, *The American Sociologist*, 2(1), pp. 92–106.

Cox, L. & Fominaya, C. F. (2009) Movement knowledge: What do we know, how do we create knowledge and what do we do with it?, *Interface: A Journal for and About Social Movements*, 1(1), pp. 1–20.

Croteau, D. (2005) Which side are you on? The tension between movement scholarship and activism, in: D. Croteau, W. Haynes & C. Ryan (Eds) *Rhyming Hope and History: Activists, Academics and Social Movement Scholarship*, pp. 20–40 (Minneapolis, London: University of Minnesota Press).

Croteau, D., Haynes, W. & Ryan, C. (Eds) (2005) *Rhyming Hope and History: Activists Academics and Social Movement Scholarship* (Minneapolis, London: University of Minnesota Press).

Davies, R. (2011) SACP: Statement by Rob Davies, South African communist party central committee member, on the Industrial Policy Action Plan, Last accessed at www.polity.org.za/article/sacp-statement-by-rob-davies-south-africancommunistpartycentral-committee-member-on-the-industrial-policy-action-plan-05052011-2011-05-05 on 20 May 2011.

Dawson, M. C. (2010) Resistance and repression: Policing protest in post-apartheid South Africa, in: J. Handmaker & R. Berkhout (Eds) *Mobilising Social Justice in South Africa: Perspectives from Researchers and Practitioners*, pp. 101–136 (The Hague: ISS and Hivos).

Desai, A. (2006) Vans, Autos, Kombis and the Drivers of Social Movements. Harold Wolpe memorial lecture hosted by the Centre for Civil Society (UKZN), International Convention Centre, July 28.

Dwyer, P. (2009) South Africa under the ANC: Still bound to the chains of exploitation, in: L. Zeilig (Ed.) *Class Struggle and Resistance in Africa*, pp. 187–211 (Cheltenham: New Clarion Press).

Fals-Borda, O. & Anisur Rahman, M. (1991) *Action and Knowledge: Breaking the Monopoly of Participatory Action-Research* (London and New York: The Apex Press).

Flacks, R. (2005) The question of relevance in social movement studies, in: *Rhyming Hope and History: Activists, Academics and Social Movement Scholarship*, pp. 3–19 (Minneapolis, London: University of Minnesota Press).

Focus Group, Ga-Chaba, Limpopo, 26 September 2009, Translated from Pedi.

Freire, P. (1972) *Pegagogy of the Oppressed* (New York: Herder and Herder).

Gibson, N. (Ed.) (2006) *Challenging Hegemony: Social Movements and the Quest for a New Humanism in Post-Apartheid South Africa* (Trenton, NJ/Asmara, Eritrea: Africa World Press).

Hamel, J. (1998) The positions of Pierre Bourdieu and Alain Touraine respecting qualitative methods, *British Journal of Sociology*, 49, pp. 1–19.

Juris, J. S. (2007) Practicing militant ethnography, in: S. Shukaitis & D. Graeber with Erika Biddle (Eds) *Constituent Imaginations: Militant Investigations Collective Theorization*, pp. 164–176 (Oakland, Edinburgh, West Virginia: AK Press).

Juris, J. (2008) *Networking Futures: The Movements Against Corporate Globalization* (Durham and London: Duke University Press).

Kindon, S., Pain, R. & Kesby, M. (2007) Introduction: Connecting people, participation and place, in: S. Kindon, R. Pain & M. Kesby (Eds) *Participatory Action Research Approaches and Methods: Connecting People, Participation and Place*, pp. 1–6 (New York: Routledge).

Marais, H. (2001) *South Africa: Limits to change—The Political Economy of Transition* (London, New York/ Cape Town: Zed Books/University of Cape Town Press).

Marais, H. (2011) *Pushed to the Limit: The Political Economy of Change* (Cape Town: UCT Press).

Mbali, C. (2010) Against journal articles for measuring value in university output, *South African Journal of Higher Education*, 24(5), pp. 745–757.

Mbali, C. (2011) Publish or be damned, mail and guardian online, Last accessed at http://mg.co.za/article/ 2011-02-25-publish-or-be-damned on 10 October 2011.

McKinley, D. (2003) The political economy of the rise of social movements in South Africa, Seminar paper presented at the Centre for Policy Studies, Johannesburg.

McKinley, D. (2006a) Democracy and social movements in South Africa, in: V. Padyachee (Ed.) *The Development Decade? Economic and Social Change in South Africa, 1994–2004*, pp. 413–426 (Cape Town: HSRC Press).

McKinley, D. (2006b) South Africa's third local government elections and the institutionalisation of 'Low-Intensity' Neo-Liberal Democracy, in: J. Minnie (Ed.) *Outside the Ballot Box: Preconditions for Elections in Southern Africa 2005/6*, pp. 151–163 (Windhoek: Media Institute of Southern Africa).

McKinley, D. (11 January 2011) Personal communication.

Pithouse, R. (Ed.) (2006) *Asinimali: University Struggles in Post-Apartheid South Africa* (Trenton, NJ/Asmara, Eritrea: Africa World Press).

Poni, M., Interview, Khayelitsha, Cape Town, 10 September 2009.

Robins, S. (2008) *From Revolution to Rights in South Africa: Social Movements, NGOs and Popular Politics after Apartheid* (Pietermaritzburg: James Currey).

Schurman, R. & Munro, W. (2006) Ideas, thinkers and social networks: The process of grievance construction in the anti-genetic engineering movement, *Theory and Society*, 35, pp. 1–38.

Seekings, J. & Nattrass, N. (2006) *Class, Race, and Inequality in South Africa* (Scottsville: University of KwaZulu-Natal Press).

Touraine, A. (1980) The voice and the eye: On the relationship between actors and analysis, *Political Psychology*, 2(1), pp. 3–14.

Touraine, A. (1981) *The Voice and the Eye: An Analysis of Social Movements* (Cambridge: Cambridge University Press).

University of Johannesburg (n.d.) DHET publication subsidy, Last accessed at http://www.uj.ac.za/EN/Research/ Research%20Information/IncentivesforResearchers/Pages/DoEPublicationSubsidy.aspx on 101 October 2011.

Walsh, S. (2008) Uncomfortable collaborations: Contesting constructions of the 'poor' in South Africa, *Review of African Political Economy*, 116, pp. 255–279.

The Gaza Freedom Flotilla: Human Rights, Activism and Academic Neutrality

ANNE DE JONG

Department of Anthropology, School of Oriental and African Studies, London, UK

ABSTRACT   *This article addresses two ethical challenges that, over the past decade, have become particularly prominent for any scholar conducting fieldwork research in contested spaces or on contested research themes. These are, first, the role researchers choose to adopt in the field and, second, the ways in which research is theoretically positioned. This article contributes to these debates by looking at binary constructs in Israel and the Occupied Palestinian Territories and the consequences on both substantive analysis and claims of academic neutrality. I will propose that theoretical positioning and the role of a researcher are not separate aspects within the ethics debate but instead should be approached as a dynamic process which requires continuous critical reflexivity. Discussing the political discourse of joint Palestinian and Israeli nonviolent activists, I argue that the participants of the nonviolent struggle do not 'merely' strive for peace but rather aim to transform the perception of the current situation from binary conflict into a 'classic' human rights struggle. Through a brief genealogy of writings on Israel–Palestine I will connect the practical positioning of the activists to academic analyses. Depicting the contemporary situation in Israel and the Occupied Palestinian Territories as 'binary conflict' with the desired solution 'peace' is not academically neutral but rather entails a paradigm that encourages binary categories which are a poor reflection of the reality, distort unequal power relations and ignores the lived experience of violence. The far-reaching consequences of the peace and conflict paradigm on academic analyses are then illustrated through a critical exploration of the 'war of narratives' surrounding the 2010 Gaza Freedom Flotilla. In conclusion, I will connect the ethics of theoretical positioning and the role of a researcher by elaborating on my personal motivation to join the Gaza Freedom Flotilla. I demonstrate how my research experience shaped my theoretical framework and how my theoretical framework subsequently significantly altered my perception of the appropriate role of a researcher. In other words, how my research results directly led to the conscious decision to join the Gaza Freedom Flotilla as an activist exactly because I am dedicated to academic research.*

Approaching Israel–Palestine

When conducting fieldwork research in a contested place (Israel and the Occupied Palestinian Territories), under contested circumstances (military occupation), focusing on a contested research theme (joint nonviolent activism and resistance) one is faced with a variety of ethical considerations. The most immediate, visible and perhaps the most personal of such questions is the threat of physical violence that accompanies conducting 'fieldwork under fire' (Robben & Nordstrom, 1995). To carry out a particular research in the Occupied Palestinian Territories one has to be willing to undergo direct interpersonal violence by the Israeli Defence Force (IDF) and settlers (Swedenburg, 1989; Bourgois, 1991). This risk intensifies if one focuses on (nonviolent) social movements and dissent because, as an anthropologist, this requires participant observation of direct actions. Despite the conscious nonviolent conduct of all participants of my research organizations, nonviolent demonstrations are regularly met by tear gas, stun grenades, rubber-coated steel bullets and even live ammunition (Allen, 2002, 2008; Carey & Shainin, 2002; King, 2007; Shulman, 2007).

One ethical consideration that has received growing attention in the last decade is the political positioning of a researcher towards his or her research interlocutors, area and theme (Hodkinson, 2004). Within the epistemic community (Caplan, 2003; De Jong, 2005) of scholars working on Israel and the Occupied Palestinian Territories this debate revolves around two main focusing points; academic or theoretical positioning and the position of the researcher (Jean-Klein, 2001; Anderson, 2002; Brown, 2007). The latter interdisciplinary, in-depth and at times heated debate concerns the role and involvement of a researcher with his or her research interlocutors and community. This includes arguments for 'militant research' (Shukaitis et al., 2007), 'embedded research' (Hodkinson & Chatterton, 2006), 'Thirdspace positioning' (Anderson, 2002) and the responsibility of researchers to be 'critical collaborators' (Routledge, 1996, 2003).

The second string of reasoning within the positioning debate focuses on theoretically approaching Israel–Palestine as a conceptualized research theme. That is, it concerns the choices of theoretical framework and conceptualizations of '[. . .] thinking Palestine; the Palestinian conception of Palestine and the representation of Palestine, the consciousness, the political idea, the territory, the history, seen, above all, as a dialectic experience positioned against its perennial other, Zionism' (Lentin, 2008, p. 2). Agamben's concept of 'state of exception' (1998, 2005) sparked a scholarly debate that aims to '[. . .] disrupt the assumptions and thinking of the powerful' (Hodkinson, 2004, p. 24). As such, approaching Israel as a racial state *par excellence* (Lentin, 2008) or as a *mukhabarat* state (Pappe, 2008) provided innovative scholarship and conceptualizations like 'spaciocide' (Hanafi, 2005, 2009; Barclay, 2010), 'thanapolitics' (Ghanim, 2008), and 'racial Palestinianization' (Goldberg 2005, 2008).

The above two streams of debate are often dealt with as separate issues (Brown, 2007) in which a researcher is supposed to 'choose' a position prior to conducting fieldwork research (Anderson, 2002; Shukaitis et al., 2007). I argue, however, that the ethics regarding positioning should be approached as a continuous dynamic process in which a theoretical approach influences a researcher's position which in turn may influence the theoretical approach. This is by no means a linear process and thus requires continuous and critical reflexivity. This argument will be clarified and deepened through a critical exposition of how my own research experience shaped my research data and concurrent theoretical analyses and how these in turn altered my perception of research position and ultimately led to the decision to join the Gaza Freedom Flotilla.

I will first look at the political discourse of joint nonviolent activists in Israel and the Occupied Palestinian Territories and show that the participants of this struggle do not 'merely' strive for peace but rather aim to transform the perception of the contemporary situation from binary conflict into a classic human rights struggle.

Subsequently, I will connect this practical positioning of the activists to academic analyses by briefly exploring the genealogy of writings on Israel–Palestine. I will argue that the largely unquestioned perception of the contemporary situation in Israel and the Occupied Palestinian Territories as 'binary conflict' with the desired solution 'peace' between the 'two sides' entails a paradigm that encourages binary categories which hold little reflection on the ground; distorts (unequal) power relations, and; sanitizes the ontics—lived experience—of violence.

I will elaborate on how the experience of fieldwork among activists and the research conclusions that sprang from this period led to my conscious decision to 'go native' and join the 2010 Gaza Freedom Flotilla. By carefully examining the discursive struggle, or 'war of narratives,' surrounding this event and its aftermath, I suggest that the Gaza Freedom Flotilla can be interpreted as a microcosm of the broader joint Palestinian–Israeli nonviolent struggle. This exposition will focus on the guiding mechanisms of the peace and conflict paradigm and how they (unconsciously) reduce nonviolent action to either ideology or mere protest. In conclusion I will pose some critical questions regarding research ethics in conflicted spaces and the ambiguous relationship between academic neutrality and human rights activism.

Joint Nonviolent Activism and Resistance

In the contemporary Occupied Palestinian Territories and Israel, including besieged Gaza and annexed East Jerusalem, there is a vibrant activist scene varying from *ad hoc* non-hierarchical grassroots initiatives against the Separation Wall to highly structured dialogue groups that, through their structure and financial flows, can be considered part of the infamous 'peace business' (Brynen, 2000). Amidst this diffused yet loosely tied activist community, one can observe 'joint nonviolent activists'. Spread over half a dozen organizations that were established in the first two years of—and as a direct response to—the outbreak of the al-Aqsa [second] Intifada (2000–2002), these Palestinian and Israeli activists consciously work together in a nonviolent manner in order to 'break the cycle of violence' and by doing so aim to undermine rigid ethno-nationalist politics that are perceived to enable and sustain the Occupation (Conner, 1994; Carey & Shainin, 2002; Godfrey-Goldstein, 2005).

While this self-declared collective[1] is regularly analysed as part of the Israeli or Palestinian 'peace camp' (Bar-on, 1996; Kaminer, 1996; Carey & Shainin, 2002), the participants of the joint nonviolent struggle do not strive for peace. That is, they do not actively use 'peace' in their terminology, do not consider themselves peace activists and even distant themselves from other so called 'peace groups'. In the words of a female Israeli activist associated with the grassroots direct action group Ta'ayush:[2]

> You can call me a peace activist if you must but our work is not about peace. Nothing here is about peace. Peace is an empty shell used by everybody. Bush claims that what he is doing in Iraq is for peace. [...] We work to end the occupation. We want to break the cycle of violence, break the system of power and simply end the occupation.
>
> (Nurit, Isr, F, Beer Sheva, 23-03-2008)

In the first year of my fieldwork research I figured that this 'problem with peace' was solely and only because the activists connect it to the failed Oslo Accords, Camp David and other peace talks perceived as 'empty' (Brynen, 2000; Pappe, 2004; Said, 2004). While the activists surely oppose negotiations in which issues they prioritize, such as refugees and the status of Jerusalem, are left unaddressed (Hashem Talhami, 2003; Fischbach, 2006), it took me some time to realize that their objection to the usage of the term 'peace' stems from its presupposed content and is directly related to the political discourse of joint nonviolent initiatives.

Joint nonviolent initiatives uphold a sophisticated, well developed political discourse based on their perception of both the contemporary situation in Israel and the Occupied Palestinian Territories and particular historical developments. All four joint nonviolent organizations central to this research[3] were established between 1996 and 2002 as a direct response to the collapse of the Oslo Accords and the outbreak of the second Intifada. This period is most often analysed with the focus on the Oslo Accords but if one looks at the consequences of this 5-year period on popular protest, two intertwined and simultaneous developments should be noticed.

On the one hand, the aftermath of the failed Oslo Accords and the assassination of late Prime Minister Yitzak Rabin combined with the outbreak of the second Intifada saw rise to the widespread perception among Israelis that there is 'no [Palestinian] partner for peace' (Barak, 2000). This perception replaced the previous dominant 'land for peace discourse'—exchanging land in return for security/peace—and has been considered the final deathblow to the Israeli left or Israeli peace camp (Avnery, 2007; Baskin, 2009; Hoffman, 2009). On the other hand, tightened occupational measures such as increased checkpoints and an extended identification system (Gordon, 2008) saw a deep scepticism among Palestinians towards the so called peace process. By 1997, 67.1% of the Palestinians believed that the Oslo Accords had 'negative' or 'very negative' effects on their economy (UNESCO, 1997). This led to the widespread perception among Palestinians that the immediate period after the Oslo Accords consisted of a continuation of occupation be it 'in a different form' (Baroud, 2006; Asad, 2007; Said, 2004). Both intertwined developments, whether real or imagined, led to increased violence and separation/segregation measures (Halper, 2000; Said, 2004; Gordon, 2008).

In response to these particular socio-economic, political and historic surroundings, Palestinian and Israeli activists developed a counter-strategy of *joint nonviolent* resistance and activism. Both nonviolent resistance and joint Palestinian–Israeli activism are not new or recent phenomena. Studies as conducted by Khalidi (1979) and Ted Swedenburgh (1995) trace this particular form of opposition back to as early as 1902. What sets contemporary joint nonviolent activism and resistance apart is their conscious stance *against* peace and *for* justice (Kassis, 2004; Zaru, 2008). That is, the participants of the joint nonviolent struggle not 'merely' strive for peace but rather aim to transform the perception of the contemporary situation from binary conflict into a 'classic' human rights struggle. In other words, the activists uphold that the contemporary situation in Israel and the Occupied Palestinian Territories is not a conflict between the East and the West, not between Arabs and Jews and not even between Israelis and Palestinians. Instead they put forward that the 'real' conflict is between those who concur to Zionist exceptionalism (ethno-nationalist politics) and those who believe that human rights should apply to all people regardless of race, religion or ethnicity (Taraki, 2006; Shulman, 2007; Qumsiyeh, 2010). This argument can be further explored by looking at academic writings on Israel–Palestine and the perceived neutrality of the intertwined concepts of 'peace and conflict'.

The Peace and Conflict Paradigm

So far I have argued that the Palestinian and Israeli activists' 'problem with peace' does not merely stem from their rejection of the Oslo Accords but rather entails a core component of their shared political discourse. The activists do not subscribe to the binary division Israeli–Palestinian/Israel–Palestine and uphold that this binary divide creates, sustains as well as forms the main ingredient of the contemporary occupation. In the writing-up period of my PhD, I analytically explored this argument by tracing the genealogy of academic writing on Israel–Palestine. By looking at the texts of early Zionist leaders such as David Ben Gurion (1886–1973) and Theodor Herzl (1860–1904), I argue that these leaders actively 'broadened the framework' in order to create the perception of binary conflict (Ben-Gurion, 1973[1967]; Meirs, 1973). This artificially created binary distinction, Arabs versus Jews, concurrently enabled a distorted perception of power relations, the legitimization of violence and the exclusion of lived experience on the ground. This can be exemplified by looking at three founding Zionist myths.[4]

First, the 'a land without people, for a people without a land' (Zangwill, 1901) founding myth denies Palestinians a national identity and instead merges them into an 'Arab entity' (Flapan, 1987; Finkelstein, 1995; Rose, 2004). This enabled the perception of a binary conflict between the little David (the Jewish people) and the giant Goliath (the Arabs) (Ben-Gurion, 1973[1967], p. 124). This not only carelessly disconnects Palestinian Arabs from any specific attachment to their land by assigning equal national value to other presumed Arab geographical locations, but also places the importance of the Jewish national claim above and beyond Palestinian and Arab claims.[5]

This paved the way for the second founding Zionist myth of 'sole peaceful intentions' or the 'Israeli self defence ethos'. This myth consists of the mainstream Zionist claim that it '[...] did not anticipate or intend resorting to force against the indigenous population to achieve its aims, but only did so as the result of an accumulation of intractable circumstances' (Finkelstein, 1995, p. 98). In other words, the Zionist movement claims it had no intention of conquest or war and held no desire to harm or expel the indigenous population in the process of realizing their ambition of creating a Jewish homeland. Regardless of the 1948 Nakba[6] and the revisionist scholarship about this period (Morris, 1987; Pappe, 2006), this claim depends fully on the binary distinction, Arab versus Jews, made by the earlier 'a land without people, for a people without land' myth. Because, only when the Palestinians as an indigenous people (or nation) are dissolved into a fusion with 'the Arabs' can one resort to the generalized and inadequate perception of an Arab Goliath versus the Jewish David and take this distortion as proof of Israel's peaceful intentions (Said, 1980; Finkelstein, 1995; Sand, 2009).

This leads to the third Zionist myth which depends on the first two and consists of the supposed Jewish choice: 'to live or perish,' that is, the supposedly zero sum choice of Israel to 'fight the Arabs' or 'cease to exist' (Jabotinsky, 1923; Shafit, 1988). This rather crude statement can only stand up if you subscribe to the view of the situation as one of conflicting nation-state claims-making (as opposed to, for example, colonial conquest) between a righteous Jewish nation with solely peaceful intentions and a hostile, non-compromising, powerful and thus Goliath-like Arab entity. If one would recognize the indigenous people as a Palestinian collective nation, the Arab opposition to the Zionist immigration stream and concurrent creation of the state of Israel can no longer be depicted in crude terms of 'Arab aggression' and Israel's show of force during the 'war of independence' can no longer be adequately captured by 'self defence' (Finkelstein, 1995; Pappe, 2006).

Early academic analyses focused solely on Jewish history and Zionist development either completely ignored the indigenous people of Palestine or portrayed them as the static and faceless 'Arab antagonist' to the only 'relevant and righteous' people of Israel: the Jews. While a few Western studies did recognize the independent existence of Palestinians and the validity of their rights and claims[7], and limited Arabic sources were available in English[8], local experiences were generally ignored and the Palestinians deemed non-existent or at best reduced to crude stereotypes of backwards barbarians. The post-1967 period saw a decline of this Zionist hegemony in Western scholarship due to an increase in the literature that included, explored and validated the Palestinian experience (Said & Hitchens, 1988, pp. 2 & 11).

The body of scholarly work dedicated to countering Zionist hegemony by including, recording and presenting Palestinian experience successfully exposed the 'land without a people for a people without a land' doctrine. By using Zionist documents and specific translations, it dismantled the Israeli self-defence ethos. By taking this body of work into account, it was no longer possible to stereotype 'the Arabs' into the aggressive enemy '[...] with the declared purpose of strangling [Israel] at birth' (Ben-Gurion, 1973[1967], p. 267). However, because this 'alternative scholarship' was presented as counter-archival and because it focused on *Palestinian* experience and *Palestinian* history, it was dismissed as the *Palestinian* 'narrative'. While I do not challenge the existence and importance of narrative in story-telling and memory, I do want to emphasize how the classification of post-1967 literature that challenged the Zionist version as the 'Palestinian narrative' enabled it to be accepted as 'scholarly knowledge' but simultaneously ignored because it presented the Palestinian (or Arab) 'side' in the conflict. While scholars interested in the Middle East acknowledged and built forward on this ever-increasing body of knowledge, Zionist academics and those interested in Israel and/or Zionist developments could and did ignore it by pointing out that it was written from 'the enemy's perspective' and thus at best questionable as a resource (Said & Hitchens, 1988; Taraki, 2006).

The unquestioned concept of plural narratives (the Palestinian narrative alongside the Israeli or Zionist narrative), and accusations of bias if both were not treated with equal weight, created a division in the mid-seventies scholarly work into three general approaches: one could focus on Jewish history or issues working within the Israeli narrative, one could focus on Palestinian experience, culture, et cetera, and work within the framework of the Palestinian narrative or one could take a more general approach and define one's research interest thematically as the Israel–Palestine conflict. To phrase one's research in terms of 'the Israel–Palestine conflict' was (and often still is) perceived as 'neutral' in the sense that it indicates an interest in the situation rather than in a particular view, population or experience.

This depiction, however, did not exempt scholars from the duty (or burden) of even-handedness. On the contrary, if one approaches a conflict in an academic matter (read: objectively), one is expected to give even space and consideration to both 'sides' of the conflict. This obsession with even-handedness and balance, brought forward within the unquestioned 'conflict paradigm,' successfully evaded pressing questions on power, practices of oppression and the nature of Israeli society and polity (Taraki, 2006, p. 453; Kovel, 2007; Said, 1995). In the early eighties, under the influence of feminist, Marxist, hermeneutic, deconstruction and cultural theory, the debate about the relation between knowledge and power opened up debates about approaching and writing on the Israel–Palestine conflict (Said & Hitchens, 1988, p. 17; Kimmerling, 2006). While this invigorated debate certainly addressed questions on suppressed narratives and gave rise to a wide range of scholarly works on the origin and continuation of the Israel–Palestine conflict with a newfound focus on power relations, knowledge and discourse,[9] it stopped short of identifying 'the conflict paradigm' (Taraki, 2006, p. 449) as a created, sustained and obstructive discourse itself.

I do not argue against the usage of 'conflict' in heuristic terms. I do argue that the perceived neutral concepts of 'peace and conflict' entail a paradigm with far reaching consequences for academic analyses (Taraki, 2006, p. 449). Just as it encouraged inaccurate interpretations of the events surrounding the establishment of the state of Israel, it deeply influences current scholarship on joint nonviolent activism and resistance and the situation in Israel and the Occupied Palestinian Territories at large. As such, the binary divisions Israel–Palestine regularly leads to the categorization of activists into being either 'pro-Palestinian' or 'pro-Israeli' (Godfrey-Goldstein, 2005; Shulman, 2007; Kaufman-Lacusta, 2010). Not only does this categorization bears no comparison with grounded experience—the participants of the nonviolent struggle position themselves towards an oppressive regime, not against one country or people—but it also distorts power relations and legitimizes violence. The construed logic of 'pro-Palestinian thus anti-Israeli' stigmatizes the wide variety of activists as 'collective enemy of the state' and thus as 'legitimate others' to whom violence may be applied.[10] Tear gas, rubber-coated steel bullets and other violent measures applied by the IDF against protesters against the Separation Wall are, for example, explained and legitimized in this manner by the claim of self-defence against a 'collective enemy.' Regardless of one's stance towards such protest and the response of the IDF to them, it should be noted that this reasoning fully depends on (1) a rigid binary division 'us versus them' (Israel vs. nonviolent protesters), (2) an active distortion in power relations that places this 'ultimate other' in stark opposition to Israel, and (3) the exclusion of analyses of the lived experience of violence.

In short, the peace and conflict paradigm encourages binary categories that do not reflect the situation on the ground; distorts (unequal) power relations, and; legitimizes and sanitizes lived experience of violence. This can be made explicit through the example of the 2010 Gaza Freedom Flotilla.

The Gaza Freedom Flotilla

Before I exemplify the consequences of the peace and conflict paradigm on the analyses of the Gaza Freedom Flotilla, I will shortly elaborate on my own participation of this direct nonviolent action as an anthropologist and an activist. As mentioned above, every researcher conducting fieldwork research in a contested space faces ethical dilemmas regarding academic positioning towards one's research interlocutors and surroundings (Anderson, 2002; Hodkinson, 2004; Hodkinson & Chatterton, 2006; Shukaitis et al., 2007). During my fieldwork research it quickly became clear that in Israel and the Occupied Palestinian Territories this is less a conscious choice of the researcher as it is a forced-upon position by outside factors and subjects (Swedenburg, 1995; Gordon, 2008).

The IDF, for example, does not differentiate between activists and a PhD student when applying tear gas during the weekly demonstrations against the Separation wall nor do settlers hold their verbal abuse when one physically distances oneself from 'the activists'. Four of my research interlocutors had noticed my initial inhibition to be classified as an activist and on my birthday presented me with a t-shirt which read 'PhD student at work: Don't shoot!'.

While surrounding institutions and subjects such as the IDF and the border control agency did not accept my self-declared outsiders' position, most activists did. Every time they introduced me to someone new, for example, they explained anthropology and anthropological research as follows: 'Do you remember the movie *Gorillas in the Mist* with Sigourney Weaver? About that woman studying Gorillas and she goes to live with them for years? Well, that is what Anne does except in her research we [the activists] are the monkeys'. The distinction 'them and I' was clear and when two interlocutors asked me to join them on the first boat to break the siege on Gaza in 2008, I courteously declined arguing that it 'would make me cross the border between being an activist and an anthropologist' (De Jong, 2009, p. 5, 2010).

In conjunction with my research results and as a direct consequence of my theoretical exposition into the peace and conflict paradigm, I gradually changed this position during the writing-up period of my PhD thesis. Following my own argument that the terminology of peace and conflict and academic analyses based on them are neither wholly neutral nor automatically independent (De Jong, 2011, pp. 66–91), I reconsidered my own position towards my interlocutors, their actions and the Palestinian–Israeli joint nonviolent struggle in general. I no longer believed that 'keeping silent' about grave human rights violations was a requirement for grounded research and I came to understand that 'participation' or 'non-participation' both constitute a political stance (Bourgois, 1991; Zinn, 2002[1994]). My fieldwork experience thus significantly altered my previous theoretical framework which subsequently changed my attitude towards both academic neutrality and the appropriate position of a researcher. When I was asked by my research interlocutors to participate in the 2010 Gaza Freedom Flotilla I decided to join for two intertwined reasons.

First, as anthropologists we enjoy the privilege of conducting long-term, in-depth fieldwork research. In my opinion, this privilege is accompanied by the responsibility to communicate one's research results to a broad audience. When one is faced with grave, structural human rights violations such as those constituting the contemporary Israeli blockade on Gaza, it is no longer enough to write academic articles and to attend conferences. I fully subscribe to the organizers' aim to 'bring humanitarian aid to the people in Gaza' and by doing so 'raise international awareness about the prison-like closure of the Gaza Strip and pressure the international community to review its sanctions policy' (Free Gaza Movement, 2010).

Second, by actively participating in the 2010 Gaza Freedom Flotilla I aimed to spark a debate among academics working on Israel–Palestine about theoretical neutrality and (unintentional) scholarly complicity. My participation can be argued to be a political act, but so is the largely unquestioned position of 'binary conflict' and its supposed even-handedness (Taraki, 2006; Qumsiyeh, 2010). My personal motivation to join the Gaza Freedom Flotilla as an academic *and* as an activist should thus be understood as a conscious re-positioning towards my research interlocutors as well as towards the theoretical conceptualizations of the peace and conflict paradigm. For these reasons I found myself aboard the *Challenger One*, the smallest vessel of the Gaza Freedom Flotilla, when it was attacked by the Israeli navy in the early hours of 31st of May 2010.

The first boat to attempt to break Israel's blockade on Gaza set sail from the Larnaca harbour in Cyprus on the 9th of August 2008. On board were 20 Palestinian, Israeli and International activists who had all been previously involved in the nonviolent struggle to end the occupation.[11] As an anthropology PhD student conducting fieldwork research on joint Palestinian and Israeli nonviolent activism and resistance I knew several of them and had followed the Free Gaza Movement from the beginning.[12] By sailing to Gaza on small boats, the activists' strove to reach two goals: (1) To physically break Israel's blockade on Gaza by sea and by doing so express direct solidarity with the people living under siege, and (2) to confront Israel's ongoing abuses of Palestinian human and political rights through citizen nonviolent, direct action, and to pressure the international community to review its sanctions policy and end its support for continued Israeli occupation.[13] Or in the words of founding member Lubna Masarwa:

> The Gaza Strip is the biggest open air prison on earth. Since 2006 Israel has subjected the Gaza Strip to an increasingly severe blockade, restricting Gaza's ability to import fuel, spare parts, and other necessary materials. The siege is illegal under international law and the United Nations representative in Gaza has called the situation a humanitarian crisis. When governments fail to act, I believe it is the duty of citizens to stand up. Through citizen nonviolent direct actions we will hold Israel accountable, stand in solidarity with the oppressed and break the siege.
>
> (Lubna Masarwa, 10-03-2010)

With these goals in mind, the Free Gaza Movement organized eight voyages between August 2008 and December 2009. On four occasions, the boats managed to break the siege and reach the shore of Gaza safely. The other times, the boats were attacked by the Israeli navy and the activists arrested. In order to strengthen the impact of their message the Free Gaza Movement decided to form a coalition with five other organizations and grassroots initiatives[14] and organize the Gaza Freedom Flotilla in 2010. The original goals and nonviolent method stayed the same but because of the size of this mission, additional safety and strategic measures were taken. As such it was decided upon that each participating organization and each individual passenger would have to sign 'points of unity' which included:

- We respect the human rights of everyone, regardless of race, gender, tribe, religion, ethnicity, nationality, sexual orientation, citizenship or language.
- The lawful inhabitants of all territories occupied by Israel since June 5, 1967, including refugees unable to return to their lawful homes in Palestine, must have unimpeded access to international waters and air space, in conformity with all UN resolutions and international law.
- We stand in solidarity with the Palestinian people, but support no particular political party or organization, without exception.[15]

In addition, passengers had to arrive in the various harbours four days before departure in order to participate in mandatory workshops and trainings. The Gaza Freedom Flotilla carried passengers with a wide variety in background, motivation and profession including many doctors, writers, politicians and university professors. In order to sustain unity and avoid confusion, passengers were trained on both the theory and practicalities of strategic nonviolent resistance, citizen's rights and duties and the legal status of Israel's blockade of the Gaza Strip. Furthermore, passengers were divided in affinity groups[16] of five to ten people and possible scenarios were extensively discussed.[17]

The various boats and ships departed from different harbours and came together at the predetermined assembly point in International waters in the afternoon of Sunday the 30th of May. Because the boats did not want to reach the territorial waters of the Gaza Strip in the dark, the steering committee of the flotilla decided to hold back and set sail for Gaza only later that evening. At this point, I found myself on board the *Challenger One*, the smallest vessel of the flotilla. The atmosphere among the passengers had changed from optimistic upon departure to tense after initial contact with the Israeli marine over the onboard radios. The message of the Israeli navy was clear: 'We will take any measures in order to enforce the blockade'. The response of the Gaza Freedom Flotilla steering committee was equally clear: 'We are an unarmed humanitarian convoy on our way to Gaza. At no point will we enter Israeli water or territory. We pose no threat to Israel or its population. We are unarmed civilians, do not use force against us. We will proceed to Gaza.'[18]

A few hours later, at approximately 4:20 am on 31st of May, Israeli navy commandos boarded the *HH Mavi Marmara* and concurrently every other vessel of the flotilla. This raid left nine passengers of the Gaza Freedom Flotilla dead and over 55 seriously injured. All boats were taken to the Israeli harbour of Ashdod where all passengers, including journalists, were held captive for a period ranging from 24 h to 5 days.[19] A full account of my experience of this raid has been published in the July–August issue of the *Middle East in London* journal (De Jong, 2010). In addition, I suggest the United Nations fact-finding mission report[20] which, in my opinion, accurately describes both my experience and broader events. For the purpose of this article, I will focus on the debate surrounding these events; in other words, on the 'war of narratives' that the boarding of the Gaza Freedom Flotilla by Israeli Navy commandos sparked. I will argue that this discursive struggle not only mirrors the broader debate on Israel–Palestine, but actually draws out the mechanisms of the peace and conflict paradigm.

The 'War of Narratives': A Microcosm of the Palestinian–Israeli Nonviolent Struggle

As set out above, the peace and conflict paradigm encourages binary divisions that hold little reflection on the ground; distorts (unequal) power relations; and, sanitizes the lived experience of violence (Taraki, 2006; Qumsiyeh, 2010). I do not argue against the usage of the terminology of peace and conflict in heuristic terms and I do not propose that conclusions drawn from this theoretical framework are automatically false or corrupted. Rather I put forward that the peace and conflict paradigm severely influences academic as well as popular analyses and should be recognized as a guiding, at times misleading, underlying mechanism. The parameters of the mechanisms of the peace and conflict paradigm can be highlighted by examining the discursive struggle surrounding the 2010 Gaza Freedom Flotilla as a microcosm of the broader joint Palestinian–Israeli nonviolent struggle.

First, it should be noted that all initial information regarding the events of 31st May 2010 came from Israeli authorities and Israeli authorities alone. The passengers as well as the journalists onboard[21] were made invisible through imprisonment for a period ranging from 24 h to 5 days. Any recorded materials from the activists and working professionals onboard of the boats were confiscated by the Israeli navy and have not been returned till this day.[22] By the time the passengers were released (and most deported), Israel already effectively created a binary space (in this case Israel vs. the Gaza Freedom Flotilla participants) in which the activists could now tell 'their side' of the story, or their 'narrative' of the event. Similarly to the earlier described creation of a Zionist versus Palestinian narrative in writings on Israel–Palestine, this created a simplified binary division in which the activists were depicted as pro-Palestinian and 'thus' anti-Israeli.[23]

In addition, Israeli authorities[24] put forward the perception that it concerned a clash between Israeli navy commandos and Turkish 'terrorists'. By bluntly ignoring and actively distorting the facts that the *HH Mavi Marmara* did not sail under Turkish flag,[25] that the Turkish government denies any involvement in its mission, that there were passengers with 31 different nationalities onboard and that the other six vessels were also entered with extreme though not lethal force,[26] Israel created the artificial binary division between Israel and Turkish aggressors. This corresponds with the manner in which Israeli authorities construct a binary division in (nonviolent) dissent. Any person currently criticizing the occupation—regardless of the argument or style of such claim—is automatically depicted as pro-Palestinian and thus anti-Israeli. This bluntly ignores the fact that the vast majority of the participants of the nonviolent struggle clearly articulate that they position themselves against ethno-nationalist politics, not against one country or one people (Shulman, 2007; Halper, 2008; De Jong, 2011).

Second, the distorted binary perception Turkish terrorist versus Israeli commandos enables the Israeli claim of legitimate violence or violence as self-defence. Regardless of one's political opinion towards this claim as an eyewitness, scholar or outside observer, it should be noted that this particular narrative of events fully depends on a distorted perception of power relations based on the previously established artificial binary division Israeli navy commandos versus Turkish terrorists. Because, if one would challenge this seeming binary division and apply a more nuanced perspective of 700 (unarmed) civilians and a fully trained, equipped maritime force, the Israeli claim of mere self-defence would at least be exposed to scrutiny. Scrutiny supported by additional data such as the autopsy report[27] would seriously, and one could argue effectively, question the death of nine and the injury of 50 more Gaza Freedom Flotilla passengers as possible self-defence. Both in the broader nonviolent struggle in Israel and the Occupied Palestinian Territories and in the specific case of the flotilla, an artificial binary division enables a distortion in the perception of power relations that in turn legitimizes the use of violence against civilians by the Israeli army and navy.

Third, the above pattern partially or completely silences the nonviolent claim of the Gaza Freedom Flotilla participants; what do the activists want? To who is their claim directed and what does it consists of (Butler, 2010)? As proposed before, the Israeli and Palestinian activists involved in the joint nonviolent struggle in Israel and the Occupied Palestinian Territories are regularly stigmatized as pro-Palestinian, thus anti-Israeli to whom violence may be applied in supposed self-defence. This reasoning is enabled and sustained by analyses based on the peace and conflict paradigm and effectively stirs attention away from the claim as put forward by the activists themselves. Though in a lesser degree, a similar process can be observed in the discursive struggle regarding the Gaza Freedom Flotilla. The binary division Israeli navy commandos versus Turkish terrorists and the subsequent claim of violence as self-defence, intentionally or not, drew the attention away (or silenced) the claim of the passengers aboard the boats; the claim that the Israeli blockade of Gaza is both illegitimate and inhumane.

A self-declared nonviolent humanitarian convoy carrying over 10,000 tons of aid and consisting of 700 passengers from over 30 different countries, can only be understood if one looks at the broader socio-economic and political surroundings. In other words, if one looks at the socio-economic, political and humanitarian reality of the sea, land and air blockade on the Gaza strip enforced by Israel. While this may seem obvious, it should be noted that the Gaza Freedom Flotilla has regularly been presented as either primarily ideologically motivated or as mere protest. Since Israel denies that there is a humanitarian crisis in Gaza, they immediately accused the passengers of being extremists and even suggested ties with well-known terrorist networks.[28] The previously criticized binary division Israel–Turkey paved the way for this reduction of the passengers to ideologists who apparently did not come to bring aid and challenge the blockade but rather came to 'attack' Israel based on a perceived pre-existing ideology.[29] Similar to the stigmatization of the Israeli–Palestinian nonviolent struggle, the disconnection of the Gaza Freedom Flotilla from its broader socio-economic and political surroundings 'simplified' this particular nonviolent action into mere ideology or symbolic protest with little space for the claim of the activists within the 'war of narratives.'

Building on this line of argumentations, the Gaza Freedom Flotilla passengers do not oppose Israel *per se* but rather position themselves towards the perceived unjust and illegitimate blockade. This is in conjunction with the broader claim of the participants of the joint Palestinian–Israeli nonviolent struggle who do not perceive the contemporary situation as a conflict between Arab–Jews, Israel–Palestine or even Israelis–Palestinians but rather between those who concur with ethno-nationalist politics and those who believe that human rights should apply to all people regardless of race, religion or ethnicity.

At this point it is important to notice that the depiction of the contemporary situation in Israel and the Occupied Palestinian Territories as a human rights struggle is not a value free description either and it can even be argued to entail a far-reaching paradigm on its own (Donelly, 1998). It is not my intention to replace one binary categorization (Israel vs. Palestine) by another (human rights vs. ethno-nationalist politics). I do advocate, however, a careful re-examination of previously perceived neutral terminology and analyses concerning Israel–Palestine. In addition, I plead for a renewed debate on research ethics in which theoretical analyses and a researcher's position are considered mutually reinforcing. Approaching ethical questions as a dynamic process may not only significantly alter our research experiences and outcomes but may also provide new insights on the ever-ambiguous relationship between academia and activism.

Conclusion

This article explored how my fieldwork experience and concurrent research results altered my perception of the appropriate role of a researcher in contested areas. By looking at the political discourse of joint nonviolent activism in Israel and the Occupied Palestinian Territories, I have shown that the activists do not 'merely' strive for peace but rather aim to transform the perception of the contemporary situation from binary conflict into a classic human rights struggle.

This practical positioning by activists led to a shift in my theoretical framework which questions the perceived academic neutrality of the terms 'peace' and 'conflict'. As such, I advance the view that depicting the contemporary situation in Israel and the Occupied Palestinian Territories as 'binary conflict' with the desired solution 'peace' is not academically neutral but instead entails a paradigm that encourages binary categories which is a poor reflection of the situation on the ground, distorts unequal power relations, and largely ignores the lived experience of violence and oppression. I clarified this argument through the example of the 2010 Gaza Freedom Flotilla. The 'war of narratives' surrounding this event not only draws out the guiding mechanism of the peace and conflict paradigm but also questions the role of a researcher when it comes to academic analyses and even-handedness.

In sum, I demonstrated that research ethics in conflict areas entail a dynamic process which requires constant and critical reflexivity. This approach draws out a variety of questions which are not limited to the epistemic community of researchers focusing on Israel–Palestine. As such, it re-opens the debate regarding human rights activism and academic neutrality. Does a certain 'sanitized description' of human rights violations equal professionalism or is it the obligation of a privileged researcher to 'go native' and represent the voice of his or her research interlocutors? Can academic analyses ever be fully disconnected from its contemporary political surroundings? And could a researcher be complicit in human rights violations by, consciously or not, reinforcing the status quo of perceptions?

This article does not aim to provide clear-cut answers to such questions. On the contrary, by looking at my own research experience I have hoped to demonstrate that this debate should be an ongoing dynamic process of critical exposition examining established theoretical perceptions through continuous reflexivity.

Notes

1. Joint nonviolent activists are spread over half a dozen different organizations and grassroots initiatives each with their own specific focus or concentrated area. Nonetheless, joint nonviolent activists classify themselves—and are recognized by others within the activist community—as a distinct group that shares a particular 'political culture' (Gordon, 2008, p. 14).
2. Ta'ayush ('living together' in Arabic) is a grassroots, non-hierarchical direct action group consisting of Palestinians and Israelis that aim to break the, in their opinion, artificial binary divide Palestinians–Israelis through direct solidarity actions such as demonstrations, working days and information distribution. For more information see: http://www.taayush.org/
3. Combatants for Peace, Ta'ayush, the Parent Circle-Bereaved Family Forum, and the Israeli Committee Against House Demolitions (ICAHD).
4. I consciously use the term 'myths' rather than 'lies' (even though many have indeed been proven factually incorrect) because as John Rose points out: '[...] a lie is "an intentionally false statement, a deliberate deception", whereas a myth is "a widely held but false notion, without necessarily deceptive intent" (Rose, 2004, p. 1)'. Whether true, false, intentional or obsolete, they form for many Israelis today an integral part of their (imagined) national history.
5. At the end of the 19th century, Arabism, the prelude to Arab nationalist thinking, brought with it a strong sense of location-specific self-awareness. In Palestine alone, there were already eight regional newspapers and 21 periodical publications. There were 98 public schools and 379 private Islamic educational facilities (Khalidi, 1979, p. 213). These figures indicate that in the period of the first Aliya, Palestine was not the backwards, stagnated country the Zionist movement would like us to believe.

6. Nakba ('catastrophe' in Arabic) refers to the events of May 1948 in which approximately 700,000 Palestinians were displaced and more than 400 Palestinian villages were destroyed in the territory that we now call Israel.

7. See for example: A.L Tibawi, 1939 *The Husain-McMahon correspondence, or: Palestine is covered by the British pledge of 1915 regarding the Arab independence.* Jaffa: Submitted to the Palestine Royal Commission. Albert Hourani, 1962 *A vision of history: Near Eastern and other essays.* Beirut: Khayats. Philip K. Hitti, 1951. *History of Syria, including Lebanon and Palestine* London: Macmillan.

8. Most notably George Antonius's *The Arab Awakening, 1938.*

9. The two main instigators of this renewed school of thought were Edward Said and Michel Foucault. While from very different perspectives, disciplines and continents and with different intentions, both emphasized the relationship between power and knowledge, taking the 'imperialist' or 'orientalist' mindset as a position of unequal, suppressed power relations not only between official doctrine and submerged discourses but also between the researcher and the researched. For further reading see: Edward Said, 1991 *Orientalism.* London: Penguin. And: Michel Foucault, 1972. *The Archaeology of Knowledge.* London: Tavistock Publications.

10. The anti-Boycott, Disinvestment and Sanctions law passed by the Knesset on the 11th of June 2011 is one example of how nonviolent protest is delegitimized by construed binary stigmatization: Lis, J. (11-07-2011) 'Israel passes law banning calls for boycott' *Ha'aretz.*

11. Passenger lists for all boat journeys can be found at the Free Gaza Movement website: http://www.freega za.org/en/all-passengers/57-first-trip-to-gaza?layout=default. Information and analyses conveyed in this article regarding the various overseas journeys to Gaza is based on organizational resources, fieldwork data (June 2007 to September 2009), personal testimonies and first-hand direct experiences (Gaza Freedom Flotilla 2011).

12. The following information stems from fieldwork notes and recorded interviews of several steering committee meetings. In addition I draw upon data retrieved through participant observation of both the preparation period and the voyage of the Gaza Freedom Flotilla itself. All activists were aware that I attended the meetings as a researcher. As will be discussed later, this changed three weeks before departure when I decided to join the Flotilla and become a passenger. All information enclosed here has been extensively discussed and approved by the activists.

13. For more information on previous voyages and the goal of the Free Gaza Movements see: http://www.freega za.org

14. The Free Gaza Flotilla coalition consists of: The Free Gaza Movement; the European Campaign to End the Siege on Gaza; IHH—the Turkish Foundation for Human Rights, Freedoms and Humanitarian Relief; the International Committee to End the Siege on Gaza; Ship to Gaza Sweden and Ship to Gaza Greece.

15. Points of Unity of the Gaza Freedom Flotilla and individual passengers. Printed with permission of the Free Gaza Movement Interim Board of Directors. http://www.freegaza.org/en/about-us/mission. Retrieved on 29-05-2010.

16. An affinity group is a small group of activists who work together on particular direct actions. Affinity groups are organized in a non-hierarchal manner and serve as a practical tool to enable consensus-based decision-making in large groups as well as to provide solidarity and emotional support to the individual activists.

17. It was agreed upon by all organizing bodies that the points of unity, passenger training and the affinity group system would apply to all passengers without exception. All the passengers that I met during the voyage and all those I have spoken to after had indeed received the training and signed the points of unity. Due to the large amount of participants, however, I cannot verify whether every single passenger aboard of the vessels met the requirements as set out by the participating organizations.

18. Abstract from communication between the Israeli marine and the Gaza Freedom Flotilla steering committee over an open line at the onboard radio: personal observation and field notes.

19. 27-09-2010 *Report of the international fact-finding mission to investigate violations of international law, including international humanitarian and human rights law, resulting from the Israeli attacks on the flotilla of ships carrying humanitarian assistance.* http://www2.ohchr.org/english/bodies/hrcouncil/docs/15session/ A.HRC.15.21_en.PDF

20. 27-09-2010 *Report of the international fact-finding mission to investigate violations of international law, including international humanitarian and human rights law, resulting from the Israeli attacks on the flotilla of ships carrying humanitarian assistance.* http://www2.ohchr.org/english/bodies/hrcouncil/docs/15session/ A.HRC.15.21_en.PDF

21. There were 60 registered journalists onboard, spread over the seven boats of the Flotilla.

22. Some personal belongings have been returned to the passengers. This did not include, however, any recorded materials of the boarding and does not include electronic equipment such as cameras, telephones and laptops.
23. The extent to which the passengers aboard the Gaza Freedom Flotilla and their sympathizers were portrayed as 'anti-Israeli' can be taken from various news reports describing erupting protests against the Flotilla raid as 'anti-Israel protests'. This included well-known news outlets such as the British Guardian newspaper ('Anti-Israel protests over flotilla attack intensify', 01-07-2010) and The Telegraph newspaper ('Gaza aid flotilla attacks: anti-Israel protests around the world', 02-07-2010).
24. The most visible spokesperson of Israeli authorities regarding the Gaza Freedom Flotilla raid can be argued to be Mark Regev. As an official IDF spokesperson he consistently subdivided the passengers aboard into 'peace activists' and Turkish terrorists; Regev, M (01-07-2010) 'Mark Regev Interviewed about Gaza Freedom Flotilla by John Snow'. *Channel 4*: http://www.youtube.com/watch?v = b06gRN_tL74. Retrieved on 05-08-2011.
25. The HH *Mavi Marmara* sailed under the flag of Comoros.
26. Serious injuries were reported among passengers of all vessels. As such, the *Challenger One* was entered with sound-grenades, paint balls and rubber-coated steel bullets. Before the Israeli commandos physically entered the boat on which I was present, I was already hit by six rubber bullets in my back followed by direct interpersonal violence including severe beating, gagging and the application of black hoods over the head by Israeli navy commandos.
27. The autopsy report concludes that at least five of the nine fatal victims of the raid on the Gaza Freedom Flotilla were shot from close range suggesting 'execution style' killings: 27-09-2010 *Report of the international fact-finding mission to investigate violations of international law, including international humanitarian and human rights law, resulting from the Israeli attacks on the flotilla of ships carrying humanitarian assistance.* Retrieved 07-08-2011 http://www2.ohchr.org/english/bodies/hrcouncil/docs/15session/A.HRC.15.21_en.PDF
28. The claim that the passengers of the flotilla either were terrorists themselves or had strong connections to registered terrorists groups has been put forward through both official IDF press releases and popular media outlets: http://www.ynetnews.com/articles/0,7340,L-3899960,00.html
29. Israeli spokespersons never fully explained why any extremists would 'attack' Israel in this manner but regularly suggested ties with Al-Qaida and thus implied that it would concern an irrational act against 'the West' that should be placed in the global 'war on terror'.

References

Allen, L. (2002) Palestinians debate 'Polite' resistance to occupation, *Middle East Report*, 225, pp. 38–43.
Allen, L. (2008) Getting by the occupation: How violence became normal during the second Palestinian Intifada, *Cultural Anthropology*, 23(3), pp. 453–487.
Anderson, J. (2002) Researching environmental resistance: Working through secondspace and thirdspace approaches, *Qualitative Research*, 2, pp. 301–322.
Asad, T. (2007) *On Suicide Bombing* (New York: Columbia University Press).
Avnery, U. (2007) *12 Years Later*, 27 October, Column Gush Shalom. Retrieved on 18-04-2009, Available at: http://zope.gush-shalom.org/home/en/channels/avnery/1193520170
Barak, E. (2000) Israel's MFT: Ministry of Foreign Affairs, The Peace Process: Key Speeches by Israeli Leaders Statement by PM Barak on his return from Camp David – July 26, 2000.
Barclay, A. (2010) Resisting spaciocide: *Notes on the spatial struggle in Israel–Palestine*, Adapted dissertation for Cardiff University.
Bar-on, M. (1996) *In Pursuit of Peace: A History of the Israeli Peace Movement* (Washington, DC: United States Institute of Peace Press).
Baroud, R. (2006) *The Second Palestinian Intifada: A Chronicle of a People's Struggle* (Ann Arbor, MI: Pluto Press).
Baskin, G. (11-11-2009) Jerusalem post: Encountering peace. The emerging bi-national reality, In: *Jerusalem Post*. Retrieved on 21-12-2010 http://fr.jpost.com/servlet/Satellite?cid=1228728209241&pagename=JPArticle%2FShowFull.
Ben-Gurion, D. (1973[1967]) *My Talks with Arab Leaders* (New York: The Third Press).

Bourgois, P. (1991) Confronting the ethics of ethnography: Lessons from fieldwork in Central America, in: F. Harrison (Ed.) *Decolonising Anthropology: Moving Further Toward an Anthropology for Liberation*, pp. 110–126 (Washington DC: American Anthropological Association).

Brown, G. (2007) Mutinous eruptions: Autonomous spaces of radical queer activism, *Environment and Planning A*, 39, pp. 2685–2698.

Brynen, R. (2000) *A Very Political Economy: Peacebuilding and Foreign Aid in the West Bank and Gaza* (Washington, DC: US Institute of Peace Press).

Butler, J. (2010) *Frames of War: When is Life Grievable* (London: Verso).

Caplan, P. (Ed.) (2003) *The Ethics of Anthropology; Dilemmas and Debates* (New York: Routledge).

Carey, R. & Shainin, J. (Eds) (2002) *The Other Israel: Voices of Refusal and Dissent* (New York: The New Press).

Connor, W. (1994) *Ethnonationalism: The Quest for Understanding* (Princeton: Princeton University Press).

De Jong, A. (2005) To exist is to resist: Non-violent resistance in the Palestinian occupied territories, Unpublished Master Thesis University of Amsterdam; available at Amsterdam University Library.

De Jong, A. (2009) The true enemy: Palestinian and Israeli nonviolent resistance, Paper presented at 'Militarism: Political, Economy, Security, Theory' conference. 14–15 May, 2009. University of Sussex.

De Jong, A. (2010) The grey zone: Academia, activism and the Freedom Flotilla, *The Middle East in London*, 7(2), July-August, pp. 4–5.

De Jong, A. (2011) The silent voice: Palestinian and Israeli nonviolent activism and resistance, PhD Thesis, School of Oriental and Africa Studies, Department of Anthropology and Sociology.

Donelly, J. (1998) *Universal Human Rights: In Theory and Practise*, 2nd ed., (New York: Cornell University Press).

Finkelstein, N. (1995) *Image and Reality of the Israel–Palestine Conflict* (London: Verso).

Fischbach, M. R. (2006) *The Peace Process and Palestinian Refugee Claim: Addressing Claims for Property, Compensation and Restitution* (Washington, DC: United States Institute for Peace).

Flapan, S. (1987) *The Birth of Israel: Myths and Realities* (New York: Pantheon).

Free Gaza Movement (2010) Gaza Freedom Flotilla mission statement, Available at: http://www.freegaza.org/en/about-us/mission Retrieved on 29-05-2010.

Ghanim, H. (2008) Thanatopoitics: The case of the colonial occupation in Palestine, in: R. Lentin (Ed.) *Thinking Palestine* (London: Zed Books).

Godfrey-Goldstein, A. (2005) Jayyous: Microcosm of the occupied west bank, *News from Within*, 21(1), pp. 19–27, Bethlehem: News from Within.

Goldberg, D. T. (2005) *The Racial State* (Oxford: Blackwell Publishers).

Goldberg, D. T. (2008) Racial Palestinianization, in: L. Lentin (Ed.) *Thinking Palestine* (London & New York: Zed Books).

Gordon, N. (2008) *Israel's Occupation* (Berkeley: University of California Press).

Halper, J. (2000) *The Key to Peace: Dismantling the Matrix of Control* (Jerusalem: ICAHD).

Halper, J. (2008) *An Israeli in Palestine: Resisting Dispossession, Redeeming Israel* (London: Pluto Press).

Hanafi, S. (2005) Spacio-cide and bio-politics: Israel's colonials project from 1947 to the wall, in: M. Sorkin (Ed.) *Against the Wall: Israel's Barrier to Peace* (New York: The New Press).

Hanafi, S. (2009) Spacio-cide: Colonial politics, invisibility and rezoning in Palestinian territory, *Contemporary Arab Affairs*, 2(1), pp. 106–111.

Hashem Talhami, G. (2003) *Palestinian Refugees: Pawns to Political Actors* (New York: Nova Science Publishers).

Hodkinson, P. (2004) Research as a form of work: Expertise, community and methodological objectivity, *British Educational Research Journal*, 30(1), pp. 9–26.

Hodkinson, S. & Chatterton, P. (2006) Autonomy in the City? Initial reflections on the social centres movement in the UK, *City*, 10, pp. 305–315.

Hoffman, C. (2009) The right way to see the left, *Jerusalem Post*, 16 April, Retrieved on 18-04-2009. Available at: http://www.jpost.com/servlet/Satellite?pagename=JPost/JPArticle/ShowFull&cid=1239710705639

Jabotinsky, V. (1923) The iron wall, *The Jewish Herald* (South Africa) Friday, 26th November.

Jean-Klein, I. (2001) Nationalism and resistance: The two faces of everyday activism in Palestine during the Intifada, *Cultural Anthropology*, 16(1), pp. 83–126.

Kaminer, R. (1996) *The Politics of Protest: The Israeli Peace Movement and the Palestinian Intifada* (Brighton: Sussex Academic Press).

Kassis, R. (2004) *The Palestinians and Justice Tourism,* Beit Sahour, Palestine: Alternative Tourism Group.

Kaufman-Lacusta, M. (2010) *Refusing to Be Enemies: Palestinian and Israeli Nonviolent Resistance to the Israeli Occupation* (New York: Ithaca Press).

Khalidi, R. (1979) *Peasant Resistance to Zionism* (Beirut: Institute for Palestine Studies).

Kimmerling, B. (2006) The continuation of the Israeli–Palestinian conflict by academic means: Reflections on the problematiques of publishing books and reviewing them, *Contemporary Sociology*, 35(5), pp. 447–449.

King, M. E. (2007) *A Quiet Revolution: The First Palestinian Intifada and Nonviolent Resistance* (New York: Nation Books).

Kovel, J. (2007) *Overcoming Zionism: Creating a Single Democratic State in Israel–Palestine* (London: Pluto Press).

Lentin, R. (Ed.) (2008) *Thinking Palestine* (London: Zed Books).

Meirs, G. (1973) *A Land of Our Own* (New York: G.P. Putnam's Sons).

Morris, B. (1987) *The Birth of the Palestinian Refugee Problem, 1947–1949* (Cambridge: Cambridge University Press).

Qumsiyeh, M. (2010) *Popular Resistance in Palestine: A History of Hope and Empowerment* (London: Pluto Press).

Pappe, I. (2004) *A History of Modern Palestine; One Land, Two People* (Cambridge: Cambridge University Press).

Pappe, I. (2006) *The Ethnic Cleansing of Palestine*. Oxford: One World.

Pappe, I. (2008) The Mukhabarat State of Israel, in: R. Lentin (Ed.) *Thinking Palestine* (London: Zed Books).

Robben, A. C. G. M. & Nordstrom, C. (Eds) (1995) *Fieldwork Under Fire: Contemporary Studies of Violence and Survival* (London: University of California Press).

Rose, J. (2004) *The Myths of Zionism* (London: Pluto Press).

Routledge, P. (1996) The third space as critical engagement, *Antipode*, 28, pp. 399–419.

Routledge, P. (2003) River of resistance: Critical collaboration and the dilemmas of power and ethics, *Ethics, Place and Environment*, 6, pp. 66–73.

Said, E. (1980) *The Question of Palestine* (New York: Time Books).

Said, E. (1995) *The Politics of Dispossession: The Struggle for Palestinian Self-Determination, 1969–1994*, p. 338 (London: Chatto and Windus).

Said, E. (2004) *From Oslo To Iraq And The Roadmap* (London: Boomsburry Publishers).

Said, E. & Hitchens, C. (Eds.) (1988) *Blaming The Victims; Spurious Scholarship and the Palestinian Question*. London: Verso.

Sand, S. (2009) *The Invention of the Jewish People* (London; New York: Verso).

Shafit, Y. (1988) *Jabotinsky and the Revisionist Movement, 1925–1948* (London: Frank Cass & Co.).

Shukaitis, S., Graeber, D. & Biddle, E. (Eds) (2007) *Constituent Imagination: Militant Investigations, Collective Theorization* (Oakland/Edinburgh: AK Press).

Shulman, D. (2007) *Dark Hope: Working for Peace in Israel and Palestine* (Chicago and London: University of Chicago Press).

Swedenburg, T. (1989) Occupational hazards: Palestine ethnography, *Cultural Anthropology*, 4(3), pp. 265–272.

Swedenburg, T. (1995) *Memories of Revolt: The 1936–1939 Rebellion and the Palestinian National Past*. Minnesota: University of Minnesota Press.

Taraki, L. (2006) Even-handedness and the Palestinian–Israeli/Israeli–Palestinian conflict, *Contemporary Sociology*, 35(5), pp. 449–453.

UNESCO (1997) Quarterly Report, April, 1997: Table 9 & 10.

Zangwill, I. (1901) Adon Olam, *The Jewish Quarterly Review*, 13(2), pp. 321–339.

Zaru, J. (2008) *Occupied with Nonviolence: A Palestinian Woman Speaks*. Minneapolis: Fortress Press.

Zinn, H. (2002[1994]) *You Can't Be Neutral on a Moving Train: A Personal History of Our Times* (New York: Beacon Press).

Sisterhood and After: Individualism, Ethics and an Oral History of the Women's Liberation Movement

MARGARETTA JOLLY*, POLLY RUSSELL** & RACHEL COHEN*

*Centre for Life History and Life Writing Research, University of Sussex, Brighton, UK, **Social Science Collections and Research, The British Library, London, UK

ABSTRACT   *In this article, we address the question of ethics in the study of social movements from the perspective of 'Sisterhood and After: The Women's Liberation Oral History Project', which will record life history interviews of 50 key activists in the UK for the British Library Sound Archive. Our research is inspired by the democratic ideals of oral historical methods and of feminism itself, yet we have discovered tensions concerning the status of individual experience and the practicalities of selection and method. Turning to other feminist scholars of women's movements we identify four broad justifications for focusing on the individual: a political understanding of the personal; situated knowledge; an investment in interview relationships and a psycho-social framework of analysis. Testing these justifications against some of the oral histories we have gathered, we conclude that they go a long way to answering the paradox of studying a movement through a few individuals' stories. But we are frank about the ethical and intellectual limits that a life history method imposes on capturing social movements. Examples from interviews with Mia Morris, Beatrix Campbell, Lesley Abdela, Ellen Malos and Juliet Mitchell will illuminate the history at stake.*

Sally Alexander [and of course others] organised the Oxford end and several of us organised the London end of Ruskin [the first 'women's liberation' conference in the UK], and several of us sent out a thousand envelopes to people all over the country, so it wasn't one person proposing it, no way was it one person proposing it. Just as I'm saying to you, I wasn't the only person thinking [about women's liberation]. These things don't happen as one person: it's a misconception of history. We get there through our different routes, because of different personal experiences, but something is happening that enables us to get there. So [. . .] if you look for people you find people, if you want to be the only heroine of the story, then you don't find them. (Juliet Mitchell, interview with Margaretta Jolly, August 2010)

Introduction

In this article, we address the question of ethics in the study of social movements from the perspective of 'Sisterhood and After: The Women's Liberation Oral History Project'.[1] This three-year project will record oral history interviews of 50 key activists in the UK for the British Library Sound Archive, as part of Leverhulme-funded research into the movement in its heyday. Our research is inspired by the democratic ideals of oral historical methods and of feminism itself. Both provide tools for answering the ethical challenges of studying a social movement. Yet, while many have seen a natural alliance between feminism and oral history's ideals, we have discovered possible tensions that echo classic debates within the movement itself.

We approach the problem from the point of view of one such debate: the status and meaning of individual experience. On the one hand, the women's movement placed extraordinary emphasis on the collective political process, and in theories of social change as grassroots-driven. Indeed, a defining ideology of the women's liberation movement, particularly in Britain, was its refusal of a 'star' system, of conventional leadership strategies and of a division of intellectual and manual labour. On the other hand, the movement believed that much of what defined the nature of gender oppression could be understood through analysing personal experience. The mantra 'the personal is political' came to reflect tensions in managing collective versus individual needs, now often summarised as the problems with 'identity politics' as they emerged in the late 1970s.

There exists an extensive literature on the study of women's social movements, including the use of life history methods within it (Taylor, 1998). Nevertheless, we have not found it easy to justify our focus on the individual experience of particular activists. The first reason is in part, this is simply because feminist methodological discussions are often rather abstract and insufficiently describe the pros and cons of particular project designs (Gottfried, 1996, p. 4). Second, it is also because oral histories of activists tend to lack information on movement size and impact, as well as on outsiders' perspectives, and ours is no exception. Third, and most ironically, social and political analysis of life history interviews is ethically difficult, in part because of the very intimacy and depth they provide (Gorelick, 1996).

We therefore begin our discussion very practically, assessing two decisions that any researcher who uses interviews faces: the selection of interviewees and the method of interviewing. Selection poses the conundrum of who can represent a process that was necessarily collective and also ideologically defined as such. How then have we negotiated the funding-driven constraint of choosing only 50 individuals to be interviewed? Second, why have we chosen to use a life history method of interviewing? Arguably this method compounds the investment in individual rather than group narratives.

In the second part, we turn to the writings of other feminists including scholars of women's movements, oral historians, narratologists and philosophers, who have sought to justify a focus on individual lives. We identify four broad justifications that grow out of feminism itself: a political understanding of the personal; situated knowledge; an investment in interview relationships and a psycho-social framework of analysis. Testing these justifications against some of the oral histories we have gathered, we conclude that they do go a long way to answering the paradox of studying a movement through a few individuals' stories. But we will be frank about the limits that our method still imposes. Before this, however, let us be equally frank about the dilemmas of selection.

The Challenge of Choosing

Whom should we interview and what resources do we have to do it? The Leverhulme Trust was generous in its funding, responding, we imagine, to our pitch that the voices of women who powered the liberation movements of the 1970s and 1980s need to be captured before it is too late. Claiming this is not difficult: it is a standard trope in oral history. And in many ways it is true. The history of feminist activism in Britain is in the paradoxical position of being both highly popular and yet done patchily and without full institutional support. So we were able to imagine that we could represent 'the movement' in a genuinely sustainable and accessible form. The interests of the archive came before our research.

But hidden in this highly ethical formulation is an obvious catch. We estimate that each of our interviews costs £650.00, if we build in time for four researchers, management across three institutions, an advisory board, transcription, travel, archival and dissemination costs. Three years seems a long time, but we now realise that a life story-based interview averaging eight hours becomes approximately 80 hours of professional arranging, summarising, transcribing and cataloguing time. Fifty interviews turns out to be all that we can fund.

We will say more in a moment about the reasons for adopting this expensive method. Let us first address the obvious and primary ethical challenge that this has left us, that of representation. How would we select from a movement which numbers conservatively in the tens of thousands? And in particular, how would we justify this not only intellectually but also in terms of honouring a movement that prided itself on its collectivity? Clearly here we are touching on a fundamental of social research, in which concerns about representativity need to be addressed for the project to be considered valid and reliable. At the same time we knew that the oral history-based nature of the British Library oral history archive and our cultural–historical frame would neither support nor require random sampling or quantitative approaches to the problem. There exist excellent discussions of the value of life history as a means to capture the deeper meanings of social protest, but we found little that directly guided us in the fundamental question of selection (Gorelick, 1996; Taylor, 1998). We could justify our eventual method as purposive sampling, but this disguises its function as a complex and ongoing negotiation between the needs of the archive, the public audience, the university and of course feminist communities themselves. We have not, for example, been able to choose 50 obscure activists, or even include any on the edge of the movement, or men, let alone anything like a control group.

We did, however, begin with the advantage that we could build on the foundation laid by the Women's Liberation Movement Research Network 2008–2009, not least because one of us, Rachel Cohen, had been its facilitator. The Network was conceived of by curators at The Women's Library, in conjunction with academics at London Metropolitan University, as a means to expand its documentation of women's campaigning history through six 'witness workshops' with women's movement activists across the UK, alongside mapping existing feminist archives.[2] These workshops, posed to correct English and London-centred biases, were held in Cardiff, Belfast, Edinburgh and Leeds as well as in London. About 240 witnesses attended in total, presenting on topics such as WLM conferences, consciousness raising, violence against women and working in trade unions, in what often became group discussions of sometimes conflicting, often nostalgic, but insightful and informative memories. The workshops were by no means representative, relying on internet advertising, the prominence of particular campaigns and friendship loops for their participants. But they do offer a fantastic first pool of women from which to pick interviewees for 'Sisterhood and After'. Further, they cover socialist, radical, liberal and national liberationist/postcolonial strands of the movement. 'Witness workshops' as a conceit implies community history, and in some respects they have also worked as a rooted form of field survey from which we have now been able to identify gaps in the record. The network was, in this respect, the first stage of a lengthy and considered, although imperfect, selection process.

Selection Criteria: Towards Representing a Movement

On this basis, we evolved a set of criteria to answer our challenge of choosing. This begins with trying to avoid the usual suspects by asking whether candidates are 'little known' or 'well known', 'already archived' or 'never archived', as well as—to put it baldly—'time running out to interview them'. Duplication and level of coverage can cut both ways, however, for we know that including a celebrity might draw the viewer to the forgotten sister beside her. And how should we respond to the Black Cultural Archives' own recently conducted oral histories of the Black women's movement?[3] We do not want to do the same job but without their insider knowledge—but nor do we want to miss out members of that crucial strand of activism. Our advisors threw in another spanner, pointing out that many activists became well known *through* the movement.

Our criteria thus also include race/ethnicity, class, religion, age, sexuality, disability, region, nation, and, separately, movement 'perspective' or 'ideology'. These categories reflect experience and in a rough way 'identity', as well as political perspective and allegiance, and certainly provide a simple map of the kinds of debates that preoccupied feminists throughout the 1970s and 1980s. However we have wished to avoid replicating either sectarian or identity politics in our selection process. For one thing, we know that interviewing someone for any length of time erodes the fixity of these categories. One example is Mia Morris' declaration in the middle of her interview that 'I have never called myself a feminist [...] We see it as a middle class, middle aged, white, exclusive club'. At the same time, she adds, 'If feminism means, you know, your right to say what you want, express what you want, have education, access, yes I'm a feminist'. Objections to the word feminist are not unusual, of course, and have been a particular point of debate in the Black women's movement since the 1970s (Helen (charles), 1997). We are still wrestling with how far to go in choosing interviewees whose movements are defined outside its terms. Morris, however, brings a valued perspective as the manager of the Black Cultural Archives women's liberation oral history. And she goes on to say that as a passionate activist she felt at odds with many black people 'who live in kind of a shell'. 'I used to feel quite lonely you know because I was involved in so much stuff and then I started to meet women in the women's movement'.[4] Beatrix Campbell's interview revealed a different kind of fluidity. Tracing her trajectory as working-class girl through to 'middle class journalist' is as complex as her story as a feminist often assumed to be a 'Geordie', though actually from Carlisle with a Dutch mother, with years of formative living in both London and Newcastle. Just as interestingly, her interview confirmed the crudity of histories that pose radical against socialist feminism. Though Campbell was a figurehead of Marxist–feminist campaigning in the 1970s (as opposed to 'socialist feminism' in her account), her thoughts on how and why she turned to issues of child sexual abuse and violence in the 1980s and 1990s reveal a more integrated view from the beginning.[5]

Thus though we have found ourselves inevitably preoccupied with identity and political allegiance, we have developed another way to select our interviewees that we hope spares us from the superficial tokenism we could so easily fall into, and that will get deeper into the historical dynamics. This is to select people primarily by their involvement in campaigns. For one thing, many campaigns and organisations are under-researched, particularly as they played out across the four UK countries and regions. But we also believe that selection by campaign involvement reflects a more material concept of history. In this sense, we reference resource mobilisation theories of social movements (Tilly, 1978). However we expand our definition of campaigns beyond public displays, or 'protest events', to include the more cultural, personal and informal forms of contention typical of women's movements during this period (Tilly, 1978; Staggenborg, 1998; Taylor & van Dyke, 2004). For example, alongside selecting Jan McKenley because of her work in the National Abortion Campaign and the Organisation of Women of Asian and African Descent (among much else), we have included Deirdre Beddoe for her work on Welsh women's history and feminist archives, and Susie Orbach for her role in the Women's Therapy Centre and her writings on body image. We are also interested in capturing memories of conferences as a specific type of women's movement activity that melds advocacy networking with lifestyle politics (Ferree & Mueller, 2004, p. 594). For instance, Lesley Abdela remembered travelling with Eastern European women on the Trans Siberian Express train to the 4th United Nations World Conference on Women's in Beijing in 1995 as a personal revelation about women's experiences of military conflict.[6] We want to define our object of analysis above all by its actions and its ideas, and to understand the lives of those who initiated, developed and fought for them.

This brings us to our second principle of selection: workplace. We have figured this by five groups: politician/public sector, academic/intellectual, grassroots/third sector, cultural activist/writer and private sector. We have done this partly to stop ourselves from simply choosing eloquent writers, though they have been formative in our own lives as women's studies graduates. Of course there is an element of cart-before-horse here, as the apparent preponderance of creatives and academics over politicians or—the category least populous to date—feminist business women, does reflect something important about the movement's cultural (but not necessarily practical) strength. Nevertheless we are here attempting to build in the potential to ask structural questions about social change and its skill-base. We know that the women's liberation movement occurred as a result of a dialectical process between ideas, grassroots activism and campaigning. But how can we best create an archive that elucidates the specifics of those dialectics, and fulfil our bold promise to our grant-body that we will offer pointers to future change-makers?

We struggle with how to create 'the best' and most inclusive shortlist we can, taking comfort from the same struggles of earlier scholars. Historian Sara Evans conducted oral histories with feminists for her important early work on the women's liberation movement, *Personal Politics: The Roots of Women's Liberation in the Civil Rights Movement and the New Left*. She, too, grappled with the problems of wanting to capture more and more stories. But Evans concluded that the substance of the argument would not be changed by more individual accounts, for the history of women's liberation activism 'is a collective story concerning patterns more than a set of individuals' (Evans, 1979, p. xi). This does not, of course, solve the political question of interview length, and whether the quality of a long interview is really worth the price. Will 50 interviews of eight hours in which a person's childhood, evolution, mid-life transitions and late life ruminations are allowed to emerge, tell us as much as 100 four-hour ones, or 400 one-hour ones? If we defend our approach, why not go even more fully towards it: why not five 80-hour interviews? Or the literary route proper: some would argue that a good biography (or auto/biography) could potentially tell as much, in the manner that, say, Barbara Taylor's recent treatment of Mary Wollstonecraft has recently done for Enlightenment feminism, or the classic of them all, Alex Hayley's oral-history based auto/biography of Malcolm X did for Black Power (X & Haley, 1968; Taylor, 2003). At issue of course, are different epistemologies as well as different methods, and whether an individual life can be recorded and analysed in a way that opens up that elusive social pattern behind protest and change. We will return to this question, but for now, let us turn to the issue of life history method. Our difficulties with selection are, we confess, partly self-imposed as the consequence of its requirements. How, then, can we justify it, intellectually, but also ethically?

Why Life Histories? The British Library and the National Life Stories Collection

Our first answer to this has to be because this is what it took for us to have the movement archived at the British Library. We could cast this as another confession of institutional conditions, but we believe that this partnership is far better described as an opportunity simply too good to refuse. The British Library's interest is undoubtedly one of the chief reasons that we can claim to have added something new to an extensive body of research, and, arguably gains over ad-hoc projects that remain within an individual researcher's attic or filing cabinet. For this project we are working closely with the library's Oral History and Social Science departments.[7] The Library's Oral History department and its fund-raising arm National Life Stories (NLS), holds one of the most importance oral history collections in the world. Moreover, it shares many of our political values. NLS was established in 1987 with a remit to fund-raise to collect oral history recordings to supplement the British Library's donated oral history collections in areas of importance for the public record.[8] Though housed in the national library in London, the NLS' collecting remit takes a broad view of history and national life and as such supplements and challenges conventional academic or national(ist) history. As the NLS website states, the charity aims to 'record first-hand experiences of as wide a cross-section of present day society as possible'.[9] Founders of NLS include Paul Thompson and Asa Briggs and the current NLS director and curator of Oral History at the British Library, Rob Perks. These three came from socialist and community oral history backgrounds, and Mary Chamberlain, a pioneer of feminist as well as socialist oral history in England, is a trustee and advisor.[10] NLS, the British Library Oral History department and the closely affiliated Oral History Society retain their commitment to supporting community oral history with a raft of public engagement programmes alongside academic partnerships.[11]

Though the library's Oral History department accepts a wide range of recordings of varying length and style, its commissioned interviews are life stories. This requires recordings that go well beyond the ostensible theme or achievement for which an individual has been selected for interview. Life story interviews are loosely chronological, starting with birth and encompassing description and reflection on family, childhood, education and work, as well as focusing in detail on an individual's contribution to a particular field or event. These interviews attempt to situate them within the broad context of their life and are designed to record an individual's biography in his or her own words. As Joanna Bornat states, the life story interview is 'more than just an extraction of information around a particular topic, it becomes an object in itself with a shape and totality given by the individual's life' (2008, p. 346). Thus, though we have developed a framework of questions and topic areas, we have been encouraged to research an interviewee's life to build awareness of significant personal as well as public trends, moments or transitions.

We have equally been encouraged to allow the recording to ebb and flow according to the interviewee's biographical narrative. While the interviewer may attempt to steer a conversation in a particular direction or may press for more details about a particular event, we must refrain from interrupting and from asking leading questions. The ideal stance for the British Library's life story interviewer is therefore not only curious, engaged, able to listen, skilled at eliciting detailed information but also able to abstain from presenting their opinion, knowledge or experience—less obviously interactive than some feminist modes of interviewing, but more reflexive than semi-structured methods. In further contrast to the average social science interview, the library's life story recordings last from six to sixty hours, though eight is the norm. The rationale is to collect as full an account of a life as possible for recordings that form part of an archive which will be accessed and analysed by future, unspecified audiences. In recalling the 450 oral histories undertaken for his 1970–1973 pioneering research into Edwardian life, Thompson retrospectively describes as a 'catastrophe' that the interviews stopped when the account reached 1918. Although the Edwardian project has been well used by subsequent researchers, information about the lives of his informants post 1918 is a lost resource (Thompson, 1996, p. 15). Thus, though we are only interviewing women who were involved at the high point of British activism—between the late 1960s and the mid-1980s—much of each interview is devoted to growing up stories about the 1940s and 1950s, and to reflections on their subsequent careers, campaigns, crises and calms, of the 1990s and 2000s.

Ethics and Life History Methods

The library's investment in a life story approach is understandable in relation to its archival function, the academic and political backgrounds of its founders and the broadening remit of a national library. But we would argue that this life story method compounds the ethical challenges faced by a typical qualitative researcher. There is an undeniable intimacy to its process—the proximity of the interviewer and interviewee, the time spent together and the intensely personal nature of the recording. The interviewer acts as a facilitator, with the quality of material they gather determined by the degree of trust and confidence an interviewee puts in them. For an interviewee, the intensity of the relationship, in fact, can obfuscate the impetus for recording and take precedence so that they might reveal more information than they retrospectively feel comfortable with. Though we follow legal and ethical guidelines in our handling of consent, and offer the interviewee the right to close some or all of their recording, this still places us in a dilemma for we are in possession of unique and exciting material but to act ethically may decide not to use it or make it available.

We also face the challenge of simultaneously working as interviewers (collecting for an archive) and researchers (working for an academic project). Though there are exceptions, in our experience oral historians pay significantly more attention to the ethical implications of the interview process than interview analysis. Where recordings are primarily being collected for archival purposes this is perhaps not surprising—the archive provides an ethical solution to the problem of analysis, delaying the process for future, unspecified audiences while softening the utilitarian drive of doing research. But academics must find a way to regard a life story interview at some level as 'data', though they have sat in an interviewee's home, accepted hospitality and have been trusted with intimate information. As difficult, they must be able to extract extracts from a recording.

As feminist researchers, a third puzzle arises. For even as we worry about respecting our interviewees, we have been questioned by other feminists as to our investment in individual rather than group narratives, potentially masking collective agency and making comparative analysis difficult. We were confronted with this recently in a meeting with researchers working on a very different study of women's social movements, FEMCIT: Gendered Citizenship in Multicultural Europe, The Impact of Contemporary Women's Movements.[12] This European Commission-funded project involves 13 partners and 42 researchers, and has attempted to assess the impact of contemporary women's movements on citizenship through several hundred interviews, with people who did and did not identify as feminist across nine countries. Interestingly, FEMCIT chose to mix quantitative methods with highly qualitative interviews in this mind-boggling four-year operation. For example, Nicky Le Feuvre's work on economic citizenship involved an analysis of statistical and bibliographical data on majority and minority women's employment; a comparative qualitative study of recruitment strategies in work–life contexts where companies have actively intervened to ensure recruitment of both ethnic minority and majority women; and a qualitative study of the work–life experience of a panel of men and women selected from the same work–life context or occupational group, using the biographical–narrative interview method (Halsaa, 2008).

The FEMCIT team were able to use both the scale and freedom of their method to pull out a typology of attitudes towards gender relations that may reveal a fascinating and 'big picture' story of slow change, reaction and absorption of feminist values in Europe. By contrast, our method evidently invests far more in our interviewees' own view of themselves, and risks compounding an embarrassing trend towards celebrity feminism. We should, of course, appreciate the bags of Euros that FEMCIT needed to fund its operation: ironically, capturing the truly unknown interviewee for the purposes of 'evidence-based' research costs 10 times as much as the already luxurious life history approach.

At this juncture, the reader might justifiably be wondering why we have not simply used the alternative model offered by community oral histories. These are defined by and for the group recorded, done either voluntarily or funded very modestly by community development or heritage grants in which the process of sharing skills, labour and results are paramount. Often, due to these conditions, interviews are short, focused on a particular time or event and used for one off exhibition or accessible publication. The fact that the interviewees and interviewers are often from the same community and that the process is understood to be a mutual learning experience means that they conform, in many ways to Michael Frisch's ideal of 'shared authority' (Frisch, 1990). Community oral histories such as the recent Heritage Lottery-funded Bolton Women's Liberation Group history and Manchester Feminist Webs, or indeed the Leverhulme Network Witness Workshops which preceded our project, embody these impressive elements and clearly emphasise the consciousness-raising element of the process in the tradition of feminist oral history from the 1970s onwards (Anderson & Jack, 1991; Armitage et al., 2003).

The obvious answer as to why we did not go this laudable route is that we are attempting to fulfil academic needs as well as the large-scale public programmes of two heavy-weight libraries which require a sustainable archive. Our funders are paying significantly more precisely for outputs more than process, though we aim for those outputs to go beyond an academic audience. It is just because of these interests that legendary feminist oral historian Sherna Gluck, recently posed the question 'Has Feminist Oral History Lost its Radical Edge?' (2011). The implication that it has become diluted by funders and academic and archival professionals is sobering and is one whose challenge goes even closer to our hearts, for our own (not always secure) careers have been undeniably helped by this project.

The Individual, the Collective and the Feminist

Sherna Gluck's timely question points out that our ethical challenge lies partly in feminism itself, as one of many post-war social movements that tried to reframe not simply the political process but the way it should be recorded historically. And yet, as we will now suggest, feminism provides its own justifications for a method that so clearly foregrounds the individual's life, perspective and experience. In this section, we will consider the writings of other scholars of women's movements, oral historians, narratologists and philosophers, to identify four justifying reasons for our method.

The first and broadest justification is that the individual's story is not necessarily individualist. This distinction is so fundamental to understanding the 'private sphere' of domestic and sexual labour that it is almost axiomatic to gender politics, expressed in the concept of the personal political, and the way that consciousness raising, coming out, speaking out and 'autobiography' played in the methods of the movement itself. What had seemed to be the individual housewife or tomboy's problem with no name, through the accumulation of 'personal' stories, became the grounds for a new conception of power relations. Of course our interviewees by definition have already 'testified'. But nevertheless we believe that a life history approach can also capture a social structure often obscured from public view. Verta Taylor positions feminist social movement studies as sharing with the women's movement itself a need to begin from 'everyday experiences of gender oppression' (1998, pp. 365–366). The powerful compatibility this offers with the British Library's wish for archival 'life stories' is no coincidence, for, as we have shown, its curators have themselves been shaped by the people's history movements of the 1960s and 1970s, seeing the everyday account as a powerful corrective to top-down theories of social change (Thompson, 1996; Perks, 2010).

Let us give the example of one of our interviewees, Ellen Malos, who we approached as a figure described as the hub of the women's liberation movement in Bristol, a catalyst for Women's Aid nationally, as well as a powerful thinker on the politics of housework. In the first two hours of recording, Malos describes some unforgettable scenes of reading Thomas Merton's (1949) *Elected Silence* as a child by the dying embers of the fire in her family's wooden house on the edge of a gold mining town in Australia; of deciding she wanted to be a monk (definitely not a nun) aged 12; of her rebellion against Methodism by Anglican flirtation; her later defiant dances with immigrant boys and curly eyeliner on a fat teenage face.

Fifteen or so years later we see her in Bristol, sadly deciding she cannot overcome her block at writing her doctorate with two young kids to care for, already having endured losing funding for her first degree in Australia because of getting married. But in 1969, around comes Lee Cataldi, another Australian and then a university lecturer in the International Marxist Group, who says 'you're sitting round on your bum'.[13] Malos goes on to tell of the ironies of being able to take on movement organising as an educated, highly politicised housewife with time now that her kids are at school. Malos' individual story thus fits existing analyses of a post-war opening up of opportunities for both working-class people and for women, with the establishment of a welfare state and greater opportunities for higher education—interestingly here in an Australian as well as British context (Heron, 1985). Her origins in a 'respectable' white working-class family, organised and morally driven, are also familiar to historians who have identified Christianity as an important stimulus to later social movement activism, though transmuted into atheism and stimulated by socialism, the new left and student movements. More complicatedly, we can see Malos' story as one in which contradictions in role propelled her politicisation: her opportunities were always cruelly mixed with closed doors.

Social movement theories are divided on precisely how political consciousness is awakened and turns into collective action and it is not our intention to debate this huge literature here (Hunt & Benford, 2004). But we hope that the interest of Malos' story as hers alone is evident, despite and because of determining structures of class, marriage, migration, motherhood or faith. We cannot offer a typology in the manner of a large-scale project like FEMCIT, but rather the paradox of the general in the particular.

The importance of the particular brings us to the second justification for our method. This is the argument put by feminist philosophers, from many different traditions, that knowledge must be grounded in material, emotional and bodily life. Nancy Hartsock's concept of standpoint theory, and a general principle of situated knowledge remain influential here (Harding, 1991; Hartsock, 1998). Our project is also inspired by feminist moral philosophy which understands social life not primarily as a march towards greater rights for the individual, but as the balancing of relationships of care and need, responsibility and vulnerability (Jaggar & Young, 1998). Tangled, yes, immersed, yes. But much truer to everyday life—even when your everyday is the dazzle of an uprising. With other social movement theorists, we believe that a close-up view of the individuals involved can reveal social changes that happen below or outside the state, crucial, in our view, to assessing both the process and the impact of women's movements (Ferree & Mueller, 2004, p. 577). What we are suggesting is that life story interviewing is well placed to reveal the emotional and physical relationships that are part of the formation of—and also consequence of—ideas, actions, politics, and indeed, the 'passions' that can produce a lifetime commitment to activism (Ferree & Mueller, 2004, p. 596). It gives time to understanding the friendships and lovers in (and against) political networks, the way that work happens at home or in unpaid, informal spheres, the way that politics is defined with and without children, sexuality and sickness, depression, art, and parents. Thus, if it is inconveniently long to do, it is a kind of homology to the way that material, sexual, relational and reproductive life is long as well.

This level of detail relates to a third reason that feminists have often favoured life history methods: their near-obsession with the integrity of the interview relationship. This has been much discussed under the rubric of 'reflexivity' and builds on the logic of situated knowledge to foreground the position and responsibility of the interviewer as well as the interviewee (Fonow & Cook, 1991). Kathryn Anderson argued in an influential early piece that 'we need to attend more to the narrator than to our own agendas', regretting the rushing of her own interviews where she 'lost opportunities for women to reflect on the activities and events they describe and to explain their terms more fully in their own words' (Anderson & Jack, 1991, p. 13). We too are finding that it takes time to 'learn to listen', just as it takes time for interviewees to talk, and our eight hours no longer seems as generous as it did when we began. This is *not* simply a matter of bringing more psychology to the exchange, though it is no accident that so many feminist oral historians have drawn on therapeutic experience. Anderson concludes that slowing down the pace is just as beneficial to the therapeutically trained interviewer who can 'gain greater awareness of how the socio-historical context can be read between the lines of a woman's "private" inner conflict' (1991, p. 23). For example, Malos attributes her later campaigns of women against violence against women, not to any personal experience of male violence but to memories of a working-class childhood, a deep identification that produced a less dramatic but pervasive sense of personal dislocation in the middle class circles of British academia. Just as important is her narrative of her tall, dark, Greek-Australian husband, whom she described as both dominating and caring, passionate about communism and labour politics and quite willing to buy a family house that could accommodate Bristol's first women's centre and refuge in the basement. His own experiences as the child of immigrants seemed oddly played out in Malos' own dislocation as a wife, following him to England, and today investigating both their global family trees after his death. What is the pattern here? One where loving feelings about the sustaining relationship of her life did not prevent a strong criticism of men as a gender-class involved in systematic abuse of women.

Psycho-social Approaches: An Ethical Solution or Another Problem?

Acknowledging psychological life is crucial, therefore, to the potential of life storying at the analytical as well as interviewing stage of oral history, but requires careful integration with socio-historical perspectives. Here we find our fourth justification, but also further ethical problems. Feminist oral historians, for example, have drawn on psychoanalysis to look at how divisions *within* the individual's account may reveal collective fantasies, fears and tensions within social movements. Historians who have used this method include Luisa Passerini (2004), who parses the homoerotic and patriarchal undercurrents of the 1968 generation in Italy through analysing interview transcripts, interwoven with accounts of her own therapeutic process, and Mary Chamberlain, whose recent work on interviews with members of the African-Caribbean diaspora looks as much at what blocks as well as enables their memories (2009). Close readings of tiny images or moments can trace unconscious as well as conscious world views (Chanfrault-Duchet, 1991). If we adopt this method with Malos' interview, we might pick out the telling image of herself as 'half a country girl' living on the edge between bush and town. Taken symbolically, we can trace this as the beginning of a personal pattern of mediation between social, academic and political worlds, a pattern that certainly reflects Malos' pragmatic and restrained temperament in the midst of passionate sectarianisms. Arguably, it can also be mined to tell us about the early movement's general inside–outside formations, its breadth, and indeed, its own collective myths of origin. Malos' striking diffidence about any claim to individual leadership reflects an important aspect of the women's liberation movement's self-image.

A psycho-social approach can therefore be worked to unpack the social significance of individual narratives, thus mitigating against their apparent individualism. Let us be candid, however, and admit that this poses another ethical problem in its turn, if we ironically disrespect an individual's privacy or their sense of their self. For example, Malos' modest interview voice belies the evidence that her interview itself gives, inadvertently, of her personally outstanding contribution as both organiser and thinker. This contradiction in itself tells of the level at which the movement's anti-individualism was ideological rather than necessarily factual. But how far is it possible for us to say this, let alone less flattering analyses of interviewee's internal inconsistencies? For this reason, oral historians have traditionally held back from some of the more explicitly therapeutic techniques of the Biographic Narrative Method, believing that the interviewee's unconscious (however collectively framed) is their business (Bornat, 2008). We walk this line as best we can, knowing in part that our particular interviewees have usually been there before us, but it is not easy.

In this light, let us briefly turn to another interviewee, Juliet Mitchell, who has been formative in situating gender psychology as integral to women's liberation, after she realised her first work, *Women: The Longest Revolution* (1966) had omitted to appreciate this. In her published work, Mitchell has also freely drawn on psychoanalytic modes of collective psychology to address the women's liberation movement's own self-image, the concept of 'sisterhood'. In a brief introduction to the first of three collections edited with sociologist, Ann Oakley, Mitchell in 1976 pointed out that 'the ideological argument for sisterhood is that sisters must be exempt from the pernicious influence of the nuclear family'. The article also argues that sisterhood must 'repeat the terms of women's social relationships with one another—relationships that, as with all relationships in a family-oriented society, are based upon a family situation' (Mitchell & Oakley, 1976, pp. 12–13). The rivalry she wished to acknowledge takes a different turn in 'Twenty Years of Feminism', a piece Mitchell wrote 10 years later (Mitchell & Oakley, 1986). Here she considered the 'unconscious function of the women's movement' to be its complicity with a longer term change in capitalism that set middle-class women's new employment against working-class women and men's redundancy. This materialist version of a political 'unconscious' clearly gets away from the more psychological take of her earlier piece. But fascinatingly, Mitchell has recently returned to the model of sisterhood as a key to understanding the contemporary as well as the 1970s' generation of women's liberation, linking it to a global demographic transition to non-procreative populations (2007). For Mitchell, the possibility of the complete technological separation of sexuality from reproduction, even though it has been as much the effect of industrialisation as of feminism, has allowed new collective imaginaries to emerge, a 'gender' imaginary which she argues can best be understood through studying sibling rivalries and alliances.

Mitchell's own fascinating attempts at theorising from inwards to outwards pose a particular problem and opportunity, in relation to my own interview with her. For though she affirmed many of her ideas, her personal story told us something different about gender psychology and sisterhood. In it, she describes her well-known decision to leave a dazzling academic career to work as a classical psychoanalyst from the late 1970s, to marry and raise a child. Of her own history she tells how her American father, with whom she had had constant contact but with whom she had not lived, died when she was 11. Her socialist–anarchist mother married a survivor of four and a half years in concentration camps and although she and her younger brother were very pleased and excited, it also felt as though the Shoah had walked through the front door of their international, extended household. An unconventional family was fine in the 1940s and 1960s but not in the 1950s of Betty

Friedan's 'feminine mystique' (2010). The establishment of a Centre for Gender Studies at the University of Cambridge, where she is now based, is something that she considered her interview must not omit to record, but it was her own experience of psychoanalysis by Enid Balint that she identified as her personal turning point. Finally, if we had to identify one motif in her narrative, it would be her memories of the progressive King Alfred's school, which she 'absolutely lov[ed] from the very first moment', and where she had her entire schooling; it seems to have been the source of her personal confidence and political imagination.

We can trace the intellectual genesis and evolution of Mitchell's feminism in this experience. However a psycho-historical approach might observe a more ambiguous relationship to both the sibling-model of sisterhood, and to motherhood. The horizontal community of school seems to be refound in activism: peer relationships that have nurtured her own feminist politics and, again let us admit, her exceptional intellectual leadership within the movement. We can also trace them in her analytic group work, which she described in interview as something she loved too, and her preference for Balint's methods. At the same time, many of the scenes she described remained within a particular circle of intelligentsia, in which black feminists' criticism of white feminists' family metaphors seem pertinent (Bhavnani & Coulson, 1986). Further, Mitchell's distinction between sexually and reproductively identified women breaks down, as her own rebirth through psychoanalysis depends on the maternal figure of Balint and the importance of her own daughter. Doesn't feminism in any case depend on biological (even heterosexual) reproduction and legacy as well as on metaphorical, educative transmission? Her proud enumeration of her daughter's own campaigning would suggest so. Mitchell's theory produces a rather conservative idea of mothering, drawing on Freudian psychoanalytic frameworks of sexual reproduction, but she herself had a radical mother and by all accounts has a radical daughter.

The symbolic crossovers and contradictions in the life of one of the most brilliant of the women's liberation movement invites us to think further about these crossovers at large. On one level, we hope that they demonstrate the power of psychologically oriented approaches to understanding social movements, building on Mitchell's own pioneering arguments that the longest revolution can only be won if the interior structure of gender changes. On another, our analysis of Mitchell's story shows how ethically as well as technically difficult it is to bring psycho-social methods to bear, even in an interview with a committed psychoanalyst. For this reason, we have sent this piece to all the interviewees it mentions. We have changed not much—but a little nevertheless, to respect their interpretations.

Conclusion

The deep story of an individual life restages the challenges that feminists themselves brought to the table in wanting to take the experiential into account. We cannot solve the ethical difficulties this brings, which include a disproportionate focus on particular individuals and invested rather than tested knowledge; and conversely, of a potentially manipulative psychological approach to analysis. We confess that an oral history of the women's liberation movement may not really be possible, only a set of interviews that we hope will be joined by many others. We also know that we have allegiances of our own that necessarily cloud (or fire) our work. American veterans Ros Baxandall and Linda

Gordon argue that the women's movement was the largest social movement in the history of the United States probably the world (2002, p. 414). As fellow travellers we might like it to be true, but is it? Even with the battery of feminist justifications we have outlined, the life history method will not tell us once and for all *how many there were* and *how resources were distributed*, let alone *how they were seen by or influenced outsiders*, basic and crucial questions about social movement operations. We will not get near to fundamental questions of influence and policy which need to be pursued through large-scale research into non-feminists, anti-feminists as well as institutions.

We concede that an ideal social science study, even one focused on the biographical experience of activism, would measure the before as well as the after, would capture a more strictly representative population and would provide more control comparisons (Giugni, 2004). Tighter methodology might answer some of the ethical challenges of selection we have faced. But we are creating an archive as well as an analysis, and we are measuring ourselves by the much broader features that Verta Taylor proposes define feminist research across the board: a gender perspective, accentuation of women's experiences, reflexivity, participatory methods and social action (Taylor, 1998). On this level, our method is consistent with our selection criteria: highly individualised but one which should open out from superficial categories of identity and experience to get at less choate structures of action, and of feeling as well. In many ways we have found a surprising accord with the British Library's NLS, as well as the Women's Library, not to mention the intriguing support of the Leverhulme. It is a partnership which we hope will add to understanding what happens 'after sisterhood' while preserving the diverse and inimitable voices of those who were there.

Acknowledgements

The authors thank Beatrix Campbell, Ellen Malos, Mia Morris, Juliet Mitchell and Lesley Abdela for their permission to quote from unpublished interviews conducted with them under the auspices of Sisterhood and After: The Women's Liberation Oral History Project. They also thank the Leverhulme Trust for its support and the anonymous referees for this paper. The authors would like to dedicate this article to their Advisory Board, to Sally Alexander for her personal encouragement, and to their interviewees.

Notes

1. See http://www.sussex.ac.uk/clhlwr/1-7-11.html
2. See http://www.londonmet.ac.uk/thewomenslibrary/aboutthecollections/research/wlmnetwork.cfm
3. See http://www.bcaheritage.org.uk/programme
4. Mia Morris, interview with Rachel Cohen, July 2010.
5. Beatrix Campbell, interview with Margaretta Jolly, September 2010.
6. Lesley Abdela, interview with Margaretta Jolly, March 2011.
7. The British Library's Social Science department was formed in 2007. The Social Science team work closely with colleagues across different library departments to develop the library's social science collections. For more information see http://www.bl.uk/reshelp/bldept/socsci/index.html
8. The British Library's Oral History department accepts deposits of oral history recordings but also fund-raises, under the auspices of the National Life Stories (NLS) collection, to record, collect and archive its own recordings.
9. NLS projects have included the following collections: Artists Lives; Architects Lives; Lives in Steel and An Oral History of the Post Office (http://www.bl.uk/nls)
10. Joanna Bornat, founder editor of the Oral History journal and society and a pioneering feminist oral historian has also had an important influence.

11. Community oral history projects supported by the library include the Vietnamese Oral History project and the Moroccan Memories in Britain project. Previous partnerships have included ESRC/AHRC-funded Cultures of Consumption project titled 'Manufacturing Meaning Along the Food Commodity Chain' (http://www.co nsume.bbk.ac.uk/research/jackson.html). For further information about public engagement programmes see the Oral History Society's training schedule available at: (http://www.consume.bbk.ac.uk/research/jackson. html)
12. See http://www.femcit.org/
13. Ellen Malos, interview with Margaretta Jolly, October 2010.

References

Anderson, K. & Jack, D. C. (1991) Learning to listen: Interview techniques and analyses, in: S. B. Gluck & D. Patai (Eds) *Women's Words: The Feminist Practice of Oral History*, pp. 11–26 (New York/London: Routledge).

Armitage, S., Hart, P. & Weathermon, K. (2003) *Women's Oral History: The Frontiers Reader* (Lincoln, NE/ Chesham: University of Nebraska Press/Combined Academic).

Baxandall, R. & Gordon, L. (2002) Second wave feminism, in: N. A. Hewitt (Ed.) *A Companion to American Women's History*, pp. 414–432 (Oxford: Blackwell).

Bhavnani, K.-K. & Coulson, M. (1986) Transforming socialist-feminism: The challenge of racism, *Feminist Review*, 23, pp. 81–92.

Bornat, J. (2008) Biographical methods, in: P. Alasuutari, L. Bickman & J. Brannen (Eds) *The SAGE Handbook of Social Research Methods*, pp. 343–355 (Los Angeles, CA: Sage).

Chamberlain, M. (2009) Diasporic memories: Community, individuality, and creativity: A life stories perspective, *The Oral History Review*, 2, pp. 177–187.

Chanfrault-Duchet, M.-F. (1991) Narrative structures, social models, and symbolic representation in the life story, in: S. B. Gluck & D. Patai (Eds) *Women's Words: The Feminist Practice of Oral History*, pp. 77–92 (New York/London: Routledge).

(charles), H. (1997) The language of womanism: Rethinking difference, in: H. S. Mirza (Ed.) *Black British Feminism: A Reader*, pp. 278–297 (London: Routledge).

Evans, S. M. (1979) *Personal Politics: The Roots of Women's Liberation in the Civil Rights Movement and the New Left* (New York: Knopf).

Ferree, M. M. & Mueller, C. M. (2004) Feminism and the women's movement: A global perspective, in: D. A. Snow, S. A. Soule & H. Kriesi (Eds) *The Blackwell Companion to Social Movements*, pp. 576–607 (Oxford: Blackwell).

Fonow, M. M. & Cook, J. A. (1991) *Beyond Methodology: Feminist Scholarship as Lived Research* (Bloomington, IN: Indiana University Press).

Friedan, B. (2010) *The Feminine Mystique* (London: Penguin).

Frisch, M. H. (1990) *A Shared Authority: Essays on the Craft and Meaning of Oral and Public History* (Albany, NY: State University of New York Press).

Giugni, M. G. (2004) Personal and biographical consequences, in: D. A. Snow, S. A. Soule & H. Kriesi (Eds) *The Blackwell Companion to Social Movements*, pp. 489–507 (Oxford: Blackwell).

Gluck, S. B. (2011) Has feminist oral history lost its radical/subversive edge? *Oral History*, 39(2), pp. 63–72.

Gorelick, S. (1996) Contradictions of feminist methodology, in: H. Gottfried (Ed.) *Feminism and Social Change: Bridging Theory and Practice*, pp. 23–45 (Urbana, IL: University of Illinois Press).

Gottfried, H. (1996) Engaging women's communities: Dilemmas and contradictions in feminist research, in: H. Gottfried (Ed.) *Feminism and Social Change: Bridging Theory and Practice*, pp. 1–22 (Urbana, IL: University of Illinois Press).

Halsaa, B. (2008) Draft Report on Design and Methodology. Working Paper No. 1, FEMCIT, Gendered Citizenship in Multicultural Europe: The Impact of Contemporary Women's Movements. Available at http://www.femcit.org/publications.xpl (accessed 6 June 2011).

Harding, S. (1991) Who knows? Identities and feminist epistemology, in: J. Hartman & E. Messer-Davidow (Eds) *(En)gendering Knowledge: Feminists in Academe*, 1st ed., pp. 100–115 (Knoxville, TN: University of Tennessee Press).

Hartsock, N. C. M. (1998) *The Feminist Standpoint Revisited and Other Essays* (Boulder, CO: Westview Press).

Heron, L. (Ed.) (1985) *Truth, Dare or Promise: Girls Growing Up in the Fifties* (London: Virago).

Hunt, S. A. & Benford, R. D. (2004) Collective identity, solidarity, and commitment, in: D. A. Snow, S. A. Soule & H. Kriesi (Eds) *The Blackwell Companion to Social Movements*, pp. 433–457 (Oxford: Blackwell).

Jaggar, A. M. & Young, I. M. (1998) *A Companion to Feminist Philosophy* (Malden, MA: Blackwell).

Merton, T. (1949) *Elected Silence*, [S.l.]: Hollis and Carter.

Mitchell, J. (1966) Women: The longest revolution, *New Left Review*, 1, pp. 11–37.

Mitchell, J. (2007) Procreative mothers (sexual difference) and child-free sisters (gender), in: J. Browne (Ed.) *The Future of Gender*, pp. 163–188 (Cambridge, NY: Cambridge University Press).

Mitchell, J. & Oakley, A. (1976) *The Rights and Wrongs of Women* (Harmondsworth: Penguin).

Mitchell, J. & Oakley, A. (1986) *What is Feminism* (Oxford: Basil Blackwell).

Passerini, L. (2004) *Autobiography of a Generation: Italy, 1968* (Middletown, CT: Wesleyan University Press).

Perks, R. (2010) *Oral History: A Political Act? Rob Perks in Coversation with Mary Stewart* (London: British Library National Life Stories Collection).

Staggenborg, S. (1998) Introduction to special issue on qualitative methods in the study of social movements, *Qualitative Sociology*, 21, pp. 353–355.

Taylor, V. (1998) Feminist methodology in social movements research, *Qualitative Sociology*, 21, pp. 357–379.

Taylor, B. A. (2003) *Mary Wollstonecraft and the Feminist Imagination* (Cambridge: Cambridge University Press).

Taylor, V. A. & van Dyke, N. (2004) 'Get up, stand up': Tactical repertoires of social movements, in: D. A. Snow, S. A. Soule & H. Kriesi (Eds) *The Blackwell Companion to Social Movements*, pp. 262–293 (Oxford: Blackwell).

Tilly, C. (1978) *From Mobilization to Revolution* (Reading, MA/London: Addison-Wesley).

Thompson, P. (1996) *Life Story Interview with Karen Worcman: Pioneers of Social Research* (London: British Library National Life Stories Collection and Qualidata Archive, University of Essex).

X, M. & Haley, A. (1968) *The Autobiography of Malcolm X: With the Assistance of Alex Haley*, p. 512 (Harmondsworth: Penguin Books by arrangement with Hutchinson).

Ethics, Activism and the Anti-Colonial: Social Movement Research as Resistance

ADAM GARY LEWIS

Cultural Studies, Queen's University, Kingston, Canada

ABSTRACT   *This paper expands the possibility of an anti-colonial theoretical grounding for social movement research. Anarchist and militant forms of social movement research serve as a starting point for theorizing ethically oriented and embedded social movement research, and theoretical connections with Gramsci's notion of the organic intellectual are set out. These positions are brought into conversation with an anti-colonial paradigm that recognizes colonialism as a continuing process of imposed and dominating relationships that needs to be both critiqued and resisted in theoretical and activist spaces. This makes it possible to articulate an ethical conception of (militant) activist research with an anti-colonial orientation, which stands as a holistic framework, takes into account previous and continued colonial relations, and centers Indigenous knowledges as the epistemological foundation of research that occurs within areas of continued colonization. That anti-colonial theoretical grounding is thereby shown to be a requirement of research on Indigenous lands or by those who continue to reap the benefits from systems of colonization—only by doing so can research be adequately positioned in pursuit of social justice against oppression and domination.*

With growing interest in social movements and the myriad forms of activism proliferating throughout the globe, activists and academics alike are exploring and refining various aspects of research into social movements. This research has turned, some might argue, away from research centered within university walls toward research in the street—research within social movements themselves. In this paper, I demonstrate how an anti-colonial ethic might be centered and further developed as part of social movement research that is embedded, participatory and active within social movements themselves. I argue, following the work of Indigenous and anti-colonial theorists, that colonialism is not only a historical relationship, but rather a continued process of intersecting domination and oppression that needs to be opposed, resisted and transformed now (in Canada, see Alfred & Corntassel, 2005; Barker, 2009). From this assertion of colonialism as a continuing process, I argue that we, as academics and activists committed to social justice, need to incorporate and further develop an anti-colonial analysis to expand our ethical research considerations.

This is especially true for those of us who continue to do our academic and activist work on Indigenous[1] lands (Kempf, 2010a), in centers of power and privilege that benefit from processes of colonization and who seek to stand with Indigenous peoples in solidarity against colonialism and all forms of oppression and domination.

My aim is to create a dialogue between anarchist and militant forms of social movement research alongside anti-colonial and Indigenous theory and practice—to recognize the anti-colonial gap that exists in social movement research. Indigenous theorists and activists have called for an analysis of colonialism by non-Indigenous peoples (Smith, 2004; Wilson, 2008), and thus my paper seeks to take up the call for anti-colonialism on the part of settlers engaged in social movement research. This paper argues for anti-colonialism as a component of social movement research that must be specified and committed to for an ethical activist research process, with an analysis of history, research, method, positionality, continued oppression and a commitment to ending all forms of oppression. As I explore below, anarchist and militant forms of social movement research serve as a starting point for theorizing ethically oriented and embedded social movement research. I then build upon them using anti-colonial theory and critical perspectives within Indigenous research frameworks to inform a specifically anti-colonial orientation toward activist research.

Social Movement Research: Social Change and Resistance

In this section, I turn briefly to recent literature on social movement and activist research that begins to inform an ethical activist research process and can be enhanced through an anti-colonial orientation. Specifically, I look at those perspectives that make a commitment to resisting all forms of oppression and domination, alongside embedded and reflexive practices (i.e., in institutional ethnography, Frampton *et al.*, 2006; Kinsman, 2006; militant research, Graeber & Shukaitis, 2007; and anarchist investigations, Amster *et al.*, 2009) as a starting point for developing an ethical orientation toward activist research.[2] These perspectives are perhaps one route in which to move research more broadly in an ethical direction that also must take on anti-colonial commitments.

Activist forms of social movement research, following Hale (2006, p. 97), can be defined as a:

> method through which we affirm a political alignment with an organized group of people in struggle and allow dialogue with them to shape each phase of the process, from conception of the research topic to data collection to verification and dissemination of results.

Activist research is firmly political in its orientation and has embedded within it relationships based on dialogue with participants.[3]

Research militancy is a current within activist research, explored by Graeber and Shukaitis (2007), who argue for a research practice that embeds the researcher/academic within social movements, and centers research as a continuous engagement of relationships and resistance. They locate social movements and their actions as 'incubators of new knowledge' (Graeber & Shukaitis, 2007, p. 11), asking how research improves the possibilities of political action. Militant research focuses on how political organization and militant action can help to understand and interpret the world. It points to the 'uniquely self-reflexive' nature of social movements, with their internal ability for critique, analysis and the distribution of perspectives (Juris, 2007). Activists are themselves creators of knowledge. Militant research, therefore, articulates the political nature of activist research. It reinforces the need for long-term and collaborative relationships, echoing Hale's definition above. The core of a militant research methodology is direct participation within social movements that allows for the cultivation of necessary long-term relationships. The embedded nature of militant research opens the potential for an understanding of the complexities and logics of social movements, allowing interventions into movement debates (Casas-Cortés & Cobarrubias, 2007; Gordon, 2007; Juris, 2007). Militant research locates the context of learning in the street, in the heart of action itself, with individuals taking action creating theory and knowledge themselves (Latif & Jeppesen, 2007). Militant research, therefore, suggests the formation of relationships that are essential for an ethical research practice.

Where militant investigation locates the political and resistant nature of activist research, anarchist methods share similar political orientations to methods of inquiry with an overt commitment to research in the furtherance of social justice with specific attention to structures of authority and the state (Fernandez, 2009; Routledge, 2009; see others from Amster *et al.*, 2009). Anarchist methods noted here illuminate two additional aspects required for setting the foundation of an ethical, and later anti-colonial, conception of activist research.

First, Fernandez (2009) points to a reflexive research process as a core component of an ethical research practice. Reflexivity, he suggests, is 'the ability of a person to stand back and assess the aspects of her own behavior, society, power and culture in relation to such factors as motivation and meaning' (Steir, 1991, cited in Fernandez, 2009, p. 99). This requires the questioning of objectivity and the separation between researcher and subject frequently found in dominant Western conceptions of research. 'Reflexivity', therefore, 'signals the understanding that an observer is just as much a part of the social setting, context and culture that she is trying to understand', and embraces participation and involvement in activism as one researches. It allows researchers a closeness to their 'subjects' and engages directly with their feelings, hopes and rationales for action (Fernandez, 2009, p. 99).[4]

Similarly, Routledge (2009) argues for a relational ethics of struggle to foreground subject positions and the important aspects of our dual roles as activist and academic. An ethical relational positionality, he argues, calls for 'dignity, self-determination and empowerment, while acknowledging that any collaborative "we" constitutes the performance of multiple lived worlds and an entangled web of power relations' (2009, pp. 88–89). To engage in such an ethical relation, activist researchers must embrace affinity—the sharing of common ground for struggle, and solidarity and support for those who resist. This leads ultimately, Routledge argues, to an activist academic practice that 'prioritizes grounded, embodied political action, the role of theory being to contribute to, be informed by, and be grounded in such action, in order to create and nurture mutual solidarity and collective action—yielding in the end a liberatory politics of affinity' (2009, pp. 90–91). Both Routledge and Fernandez highlight crucial aspects that must be considered when attempting to formulate an ethical research practice. We must have a commitment to others we research with, as well as a commitment to act against oppression and domination and for social change.

Finally, an activist research methodology requires the realization that the structures of the academy must also be sites of struggle and resistance. The university, Casas-Cortés and Cobarrubias (2007) argue, is part of the neoliberal (and colonial, as highlighted by Indigenous theorists below) order and reproduces elements activists seek to fight against. We must see the university as seeking to 'categorize and classify' in the maintenance of power and domination (Graeber & Shukaitis, 2007). We can take up institutional ethnographic (IE) methods to reveal the institutional constructions of power, social relations and administrative regimes, making visible their contradictions and weak points to make activism within them more effective (see Frampton *et al.*, 2006). IE is employed to map social relations to produce knowledge to change the world (Kinsman, 2006). We thus have an obligation, as activist academics, to facilitate resistance in these spaces. As Casas-Cortés and Cobarrubias (2007, p. 124) argue, the ivory towers of academia must be a space of our resistance: 'They must be laid siege to, they must be infiltrated'.

By looking at movement literature that puts forth an explicitly political project and embedded research focus, I have traced some of the basic aspects of what ethical activist research might encompass. Embeddedness, self-reflexivity, affinity, the cultivation of relationships and resistance within the academy itself are all aspects that need to be considered for the foundation of ethical activist research.

Embedded Organic Intellectuals

The role of the theorist in activist research still raises some concerns and I point to, briefly, the Gramscian 'organic intellectual' as one possible orientation for activist research. The role of the intellectual within social movements, as has been already argued, is that of critique, interpretation, theoretical expansion and elaboration of what may be called the 'collective theorizations' (Graeber & Shukaitis, 2007) of social movements. The role of intellectuals explored here presents possibilities for an ethical orientation of activist research that remains committed to social movements themselves, rather than strictly academic work, and resisting all forms of oppression and domination. Gramsci's organic intellectual elaborates the role of intellectuals in social movement-based research and reinforces the potential for the intellectual to be an embedded component to aid activist movements (Gordon, 2007). This embeddedness presents a model for activist research, noted by several theorists who draw from Gramsci: intellectuals as facilitators and clarifiers of movement strategy (Gordon, 2007), as 'movement intellectuals' (Haluza-DeLay, 2003) similarly embedded and aligned with movements and 'committed intellectuals' focused on the struggle to end all forms of exploitation (Fischman & McLaren, 2005).

Gordon (2007) suggests the role of the intellectual as a facilitator for movement theory, debate and action, as well as taking up the role of revealing assumptions and contractions. The embedded intellectual looks at the internal dynamics of the movement and seeks to theorize and clarify to provide further tools for struggle—to focus on the agency and 'unofficial thought' of activist theory formed in the context of struggle (Cox & Barker, 2002). The role of the intellectual is to analyze social relations and assist in the articulation of resistance strategy (Haluza-DeLay, 2003), and to mobilize/radicalize movement participants and outsiders (Cox & Barker, 2002). The activist intellectuals then might offer their conclusions, observations and theorizations back to movements, 'not as prescriptions', Graeber argues (2004, pp. 10–12, cited in Gordon, 2007, p. 278), 'but as contributions, possibilities—as gifts'. Movement activists, therefore, can fulfill a role of being organic intellectuals within their own movements (Kinsman, 2006), with theory rising from activism.

Gramsci's organic intellectual is one means of theorizing how intellectuals and academics might become embedded and linked in explicitly political ways to social movement struggles. These links and relationships form the base of ethical research within social movements committed to resistance against domination and oppression, moving academic work from mere theory to a tool for action. This conception, however, does not address the possibilities of solidarity-based relationships between Indigenous peoples and settlers that will be necessary for resisting colonialism. An anti-colonial analysis, I argue next, allows for a more specific ethical research orientation that takes stock of the activism and research that continues to occur on the colonized lands of Indigenous peoples, the lands of powers that reap the benefits of continued colonialism and between Indigenous and non-Indigenous/settler peoples.

Articulating the Anti-Colonial

To advance anti-colonialism as part of ethical activist research, a basic definitional framework needs to be established. Colonialism has been generally defined as 'direct territorial appropriation of another geo-political entity, combined forthright exploitation of its resources and labour, and systematic interference in the capacity of the appropriated culture … to organize its dispensations of power'. The persistence of colonialism after 'independence' struggles can continue as 'internal colonialism' where the dominants continue a similar administration of power over other groups (McClintock, 1992, p. 88). The core aspect of colonialism is the often violent maintenance of 'structural domination and a suppression … of the heterogeneity of the subjects in question' (Mohanty, 2003, p. 18). Mohanty's definition begins the move toward colonialism incorporating intersecting systems of oppression, beyond only a historical designation. Further, neo-colonialism points to the creation of new forms of colonialism and domination (Kempf, 2010b).

In part, colonialism has been historicized through the label of 'post-colonialism'. This is not to suggest post-colonial theory as irrelevant to anti-colonial struggles, but rather as a misleading term, an 'unrealistic rupture, a break, a move away from one condition to another' (Dei & Asgharzadeh, 2001, p. 304), a term promoting a hierarchy of historical epochs that centers colonialism and its European lineage as the referent period, 'prematurely celebratory' given the persistence of colonial dynamics (McClintock, 1992, p. 87). Colonialism is based on power and oppression, and as long as social relations continue to be structured as such, there will be little justification for the 'post' in post-colonialism (Dei & Asgharzadeh, 2001). Post-colonialism might then be better situated as an aspiration, a hope for a non-colonial future (Battiste, 2002). Colonialism, therefore, needs to be regarded as a 'transhistorical' phenomenon, continuing in contemporary societies, not relegated to a former past (Kempf, 2006, 2010b). Colonialism has not disappeared, and therefore it is to anti-colonialism that we must turn (Smith, 2002). Activists and researchers need to take stock of the continued presence of colonialism. To do otherwise would be to assert that specific forms of oppression and domination do not persist.

An anti-colonial orientation, therefore, can be articulated, following Dei (2006, p. 2) as an 'approach to theorizing colonial and re-colonial relations and the implications of imperial structures on the processes of knowledge production and validation, the understanding of indigeneity, and the pursuit of agency, resistance and subjective politics'.

Anti-colonialism is both a critique of colonial structures and processes, and a means by which to resist them (Dei & Asgharzadeh, 2001). Anti-colonialism, in the sense discussed here, promotes resistance to all aspects of oppression and domination, and as such is a holistic approach to resistance (Kempf, 2010a), which is more generally applicable as a strategy of resistance than approaches that focus on a single axis of oppression (i.e., class) (Dei & Asgharzadeh, 2001; Dei, 2010).[5] This is not to suggest that all anti-colonial theory takes up an intersectional understanding of oppression, but suggests that an ethically grounded anti-colonial theory would take up intersectionality to recognize the complexity of oppression and domination that occurs under colonial conditions (see Cannella & Manuelito, 2008; Barker, 2010).

Fundamentally, an anti-colonial orientation theorizes colonialism through the lens of Indigenous knowledges, epistemologies and histories of resistance in everyday life (Dei, 2006). It is an 'epistemology of the oppressed', based in localized knowledges and processes (Dei & Asgharzadeh, 2001, p. 300; Dei, 2010). It locates dominant ideologies and methodologies within racially based epistemologies bound up in European dominance (Hales, 2006), and takes a holistic approach, identifying spirituality and the connection of all things within relationships (Dei, 2006). Cannella and Manuelito (2008, p. 56) summarize the work of anti-colonialism to:

> a) reveal and actively challenge social systems, discourses and institutions that are oppressive and that perpetuate injustice ... and explore ways of making these systems obviously visible in society; b) support knowledges that have been discredited by dominant power orientations in ways that are transformative ... and c) construct activist conceptualizations of research that are critical and multiple in ways that are transparent, reflexive and collaborative.

To articulate an ethical conception of activist research, an anti-colonial orientation stands as a holistic framework, takes into account previous and continued colonial relations, and centers Indigenous knowledges as the epistemological foundation of research that occurs within areas of continued colonization. It centers oppositional frameworks to resist dominant European conceptions of research that continue to uphold colonial and dominating relations within academic spaces. It moves beyond assertions of 'impartiality, non-partisanship and indifference' within academia and maintains that discursive practices can never be apolitical (Dei & Asgharzadeh, 2001, p. 318). Epistemologically, anti-colonialism is grounded in the knowledge held by the oppressed and the drive to hold colonizers and settlers accountable to their histories and privileges accruing from colonial processes. From an axiological standpoint, it focuses on articulating strategies for resistance and social transformation. Ontologically, it recognizes change not only as desirable, but as possible, and rooted in the resistance of oppressed peoples (Kempf, 2010b). Anti-colonialism is a specifically political project aimed at resistance to domination and oppression, a project centered in Indigenous knowledges, and one that can further specify an ethical project of social movement research.

Centering Indigenous Knowledges and Research

I now turn to Indigenous knowledges and worldviews, primarily centered, though not exclusively so, in the context of Turtle Island (North America), to ground an anti-colonial perspective. I cannot claim to adequately articulate the complexity of Indigenous worldviews (nor would it be appropriate for me to do so), but rather I aim to present some common threads, drawing from recent works by Indigenous theorists on Indigenous research methodology and critiques of dominant Western forms of research (i.e., Denzin *et al.*, 2008; Wilson, 2008; Kovach, 2009). I argue that it is these Indigenous-oriented perspectives that must inform an ethical anti-colonial articulation of social movement research.

Indigenous worldviews and epistemologies are what fundamentally inform Indigenous research methodologies. Indigenous epistemologies are, in a general sense, encompassing of a holistic and dynamic approach to the world, ecologically centered in connections to land and place and fundamentally focused on interrelated relationships between all living things (Battiste, 2002; Little Bear, 2002; Kovach, 2009; Getty, 2010). An Indigenous research methodology, therefore, follows from an Indigenous epistemology and is centered in relational accountability or, following Wilson (2008), being accountable to all ones relations and ones community. Community accountability means, in this sense, that the community directs research, and the researcher seeks to privilege community needs and values over their own (see also Kovach, 2009). Research is not located as extractive, in comparison to Western 'smash and grab' practices (Smith, 2004), but as belonging to the community. The community 'owns' the knowledge and maintains control over the research processes, publication and reporting (Louis, 2007).

It must be noted, however, that an Indigenous epistemology, although possessing some general tenets, stems from a specific place-based context. Place maintains the connection between the past and present and situates the particular knowledges of specific Indigenous peoples that cannot be universalized. It is a connection to land based on 'collective responsibility and stewardship' (Kovach, 2009, p. 63). Indigenous knowledges and research are then firmly rooted within Indigenous communities, based on relationships of respect and accountability (Battiste, 2002; Kovach, 2005; Wilson, 2008).

Language is intimately connected to place and forms a core component of Indigenous epistemology and methodology. Languages are fluid in their ability to articulate relationships (Kovach, 2009), taking on a verb-focused, as opposed to subject-focused, framework reaffirming the active nature of relationships (Little Bear, 2002). Language emphasizes oral forms of communication and transfer of knowledge. Storytelling is emphasized as a means to hold and transfer knowledges and maintain their continuity from the past into the future (Kovach, 2009). It situates community members with their historical lineage and validates the lives of the people (Thomas, 2005). It is also a means to articulate multiple truths with regard to history and events, where every member of the community is a part of the larger articulation of histories and collective memory (Smith, 2004).

As this brief overview of Indigenous epistemology suggests, epistemological groundings cannot be separated from elements of research methodology. Indigenous research methodology is firmly grounded in cultural understandings and histories. It is, as Wilson (2008, p. 11) affirms, a 'ceremony' based on the desire to build stronger relationships and raise consciousness. An Indigenous research paradigm must employ values and beliefs that are valid to Indigenous communities and not necessarily valid to Western institutions or systems of thought. This does not require that an Indigenous paradigm be insulated from broader critical research perspectives, but rather recognizes that the primary aim of research is to benefit the community and contribute to resisting colonialism.

An Indigenous research methodology is, as a result, open-ended and is not necessarily constrained by a specific set of methodological prescriptions. An Indigenous methodology focuses on the importance of the journey in creating and conducting research as much as the importance of results, conclusions and publications (Kovach, 2005). It is also explicitly tied to Indigenous knowledges and their maintenance and revival as a means of resistance (Doxtater, 2004; Simpson, 2004). Louis (2007, p. 133) summarizes the four commonalities of Indigenous methodologies as follows: 'relational accountability' (recognizing dependence on all of ones relations), 'respectful representation' (how the researcher presents themselves and their research), 'reciprocal appropriation' (how one is invested in the landscape and how landscape is incorporated into experience) and 'rights and regulation' (research driven by community protocols).

Coming from a specific epistemological context, Indigenous research methodology comes into conflict with Westernized assumptions present within research and advances a specific critique of the colonial residues within Western paradigms. Smith's *Decolonizing Methodologies* (2004) critiques the imperial and colonial histories/contexts and impacts that Western research and methodologies have had on Indigenous peoples. She locates Western epistemologies as based in a view of history as a totalizing, singular, universal and linear narrative, constructed as innocent, depoliticized. Research, therefore, is constructed 'through imperial eyes':

> which assumes that Western ideas about the most fundamental things are the only ideas possible to hold, certainly the only rational ideas, and the only ideas which can make sense of the world, of reality, of social life and of human beings. (Smith, 2004, p. 56)

An Indigenous research paradigm seeks to resist the persistence of Western hegemony in research, to resist the 'jagged worldviews' that have been internalized by Indigenous peoples as a result of colonization (Little Bear, 2002). Previous Western research within Indigenous communities is identified with a lack of relevance and accountability to communities (Wilson, 2008). Indigenous methodologies, therefore, aim in part to disrupt the Western homogenization of research.

To counter the Western homogenization of research, Smith (2002) argues that Indigenous research and methodologies should be open to any theoretical perspective that can assist in the struggle against colonialism, while recognizing, as suggested above, that any methods used must maintain their relevance to Indigenous communities and their struggles against colonialism. Indigenous peoples need not outright reject the possibilities of resistance and participation in the academy, but aim to seek out allies and theoretical paradigms that will assist in their struggles. Theory is a tool, a means to 'write back' against the dominant narratives and constrictions of history and society (Smith, 2004). Kovach (2009, p. 26) locates an Indigenous methodological affinity with qualitative methods and cites Denzin and Lincoln (2003, p. 26) stressing 'the socially constructed nature of reality, the intimate relationship between the researcher and what is studied, and the situational constraints that shape inquiry'. Similar affinities can be observed with social movement research perspectives outlined earlier in this paper with regard to relationships and embeddedness. Qualitative research poses a similar emphasis on reflexivity and self-location within research as an inclusive process acknowledging multiple truths (Kovach, 2009). In particular, participatory critical theory, when grounded in local community contexts and knowledges, has the potential to be used as a tool in struggles against colonialism (Smith, 2002, 2004; Absolon & Willet, 2005; Denzin *et al.* 2008; Grande, 2008; Kovach, 2009) and an anti-colonial social movement ethic. For the employment of other research perspectives, any research related to Indigenous peoples, their histories and experiences must be compatible with an Indigenous epistemology and centered within Indigenous voices (Smith, 2004; Wilson, 2008). The caution that must be given to any degree of borrowing or synthesis is the recognition of the colonial history within which research is embedded (Kovach, 2009).

If we are to suggest an Indigenous perspective for social movement research, we must consider the implications. Kovach (2009, p. 30) suggests that the introduction of an Indigenous perspective into any other discourse 'must ethically include the influence of the colonial relationships, thereby introducing a decolonizing perspective to a critical paradigm'. Kovach centralizes the decolonizing nature of an Indigenous perspective on research as a means to center an Indigenous cultural grounding for research, decentering a focus on settlers and problematizing the colonial relationship. A decolonization agenda must be part of Indigenous research, and, as I argue, social movement research, because of colonialisms' continued influence. It serves as a means to unify Indigenous people into a common struggle, and aims to center voice and representation, identify colonial residues and promote change and transformation. Decolonization marks its focus on the contradictory experiences that have occurred under colonialism and the struggle to reclaim culture, knowledge and community (Smith, 2004). For settlers and those seeking to ally with Indigenous peoples, the project of decolonization must be a 'long-term process involving the bureaucratic, cultural, linguistic and psychological divesting of colonial power' (Smith, 2004, p. 98). It must be one where settlers recognize their privilege and power, derived from the continuation of a colonial system, and actively undertake a process of 'unsettling' (Regan, 2010) within their own communities. Decolonization is more than simply the elevation of previously colonized populations to places of power. Rather, it requires the 'reevaluation of the political, social, economic and judicial structures themselves and the development, if appropriate, of new structures that can hold and house the values and aspirations of the colonized' (Laenui, 2002, p. 155). Alongside the theoretical work of anti-colonialism cited above, Indigenous research promotes a decolonizing perspective, embedded locally in communities, that warrants attention from social movement researchers.

As social movement researchers and activists committed to social justice and transformative change, and often working for such goals in colonized lands and spaces, a strategy for resisting colonial realities must be part of our theory and practice. An anti-colonial analysis names colonialism as a system of oppression that must be opposed and recognizes that the work that occurs on Indigenous lands occurs within the context of colonial privilege and domination. An anti-colonial analysis asks social movement actors, committed to removing all forms of oppression and domination, to look at their own privileges and take up a politics of solidarity with Indigenous peoples. Anti-colonial work in practice, following Barker (2010), suggests that what is important is not simply whether or not settlers have been unsettled and made aware of colonialism and their relationship to it (although an important first step), but rather what the settler decides to do—whether they will seek to resist colonialism as an ally with Indigenous peoples or whether they will choose to do nothing. The only contention I have with Barker is his assertion that we must respect those who choose to do nothing once they are made aware of their colonial privileges (2010, p. 323). Rather being aware of privilege indicates the point where an individual has an obligation to work against such privileges and commit to a politics of decolonization. We cannot, as settlers and peoples committed to resisting all forms of oppression, let others continue colonial dynamics. Understanding our position as settlers requires us to take action and commit to a decolonizing and unsettling framework. It recognizes that 'colonialism is a narrative in which the Settler's power is the fundamental reference and assumption, inherently limiting Indigenous freedom and imposing a view of the world that is but an outcome or perspective on that power' (Alfred & Corntassel, 2005, p. 601). Part of our work as social movement researchers and activists must be taking up unsettling and decolonizing work once we recognize colonialism. An ethical orientation to activist research must, therefore, be one that recognizes colonialism as a force of oppression and domination to be resisted, and one that takes concrete steps to subvert colonial privileges and ally with Indigenous peoples in resistance through research and action.

Indigenous research paradigms present a means to resist colonial realities, ask us to look at the academy and educational institutions as sources of privilege and power, and ask how we might act ethically in these systems of colonialism (Battiste, 2002; Kovach, 2009). They ask us to consider whether or not we might be taking or making space for alternative critical research paradigms (Kovach, 2009), how we might learn from others, as opposed to about them, and how we might acknowledge privileges, histories and differences and work across them (Jones & Jenkins, 2008). They forcefully align, according to Denzin and Lincoln (2008, p. 15), 'the ethics of research with a politics of the oppressed, with a politics of resistance, hope and freedom'.

Indigenous methodologies are not alone in attempting to disrupt homogenous constructions of research. Critical social movement research methods, among others, have attempted similar disruptions. The key difference within an Indigenous methodology, grounded in localized communities and knowledges, is the explicit recognition and commitment to resist the residues of colonialism that continue to be present in research practice. This is a commitment that must be taken up by activists and activist intellectuals who continue to struggle for social change on colonized lands, and in centers of power that benefit from colonial processes. It is a commitment that calls for a process of unsettling within ourselves, our communities, our research and our practice, while standing in solidarity with Indigenous peoples.

Conclusion

Following the above discussion of social movement, anti-colonial and Indigenous research methods and theory, an anti-colonial ethic for research can be further enhanced within social movement research. The anti-colonial ethic must emerge as a requirement for social movements, theorists and activists who struggle against oppression and domination on Indigenous lands or who continue to reap the benefits from systems of colonization. Indigenous theory points to place, and our relationships that emanate from land bases, as a means to locally ground our theories and actions and to recognize and challenge the colonization of land (Grande, 2008). We, speaking from a settler activist standpoint, need to consider such connections, and how we are bound up within systems of colonialism. We must continue our ethical activist research work, maintaining embedded relationships, reflexivity and a commitment to resist oppression and domination, all aspects that resonate with Indigenous and anti-colonial articulations. But we must go further. We must recognize the persistence of colonialism in intersecting systems of oppression and domination and seek to include such an ethical understanding into our research practice. We must recognize ourselves as allies in solidarity with Indigenous and anti-colonial struggles (Max, 2005), with the imperative to unsettle and decolonize within our own communities and selves. We must rethink our collaborations, our contexts, our privileges and our practices, and conceive of them ethically in anti-colonial terms as a process that is never complete. 'Anti-colonialism', in the words of Dei (2010, p. 255), 'must be articulated in the interests of all who struggle against colonialism, racism, myriad oppression, capitalist imperialism, and other antihuman systems'. Anti-colonialism, to borrow from bell hooks, must be for everyone.

Acknowledgements

Special thanks to those who have assisted in the preparation of this paper: Richard Day, Sara Matthews, Gary Kibbins, Laura McDonald and the very helpful reviewers and editors at Social Movement Studies. I would also like to acknowledge the traditional territory of the Haudenosaunee Peoples of Six Nations, on whose land much of my work has occurred.

Notes

1. I use the word Indigenous in the sense of Alfred and Corntassel (2005, p. 597): 'The communities, clans, nations and tribes we call *Indigenous peoples* are just that: Indigenous to the lands they inhabit, in contrast to and in contention with the colonial societies and states that have spread out from Europe and other centers of empire. It is this place-based existence, along with the consciousness of being in struggle against the dispossessing and demeaning fact of colonization by foreign peoples, that fundamentally distinguishes Indigenous peoples from other peoples of the world'. Although I find myself located within the North American context of colonialism and Indigenous resistance as a white, straight, settler solidarity activist, I hope that the arguments put forth in this paper find resonance within broader movements of resistance.
2. I have selected these perspectives because (1) they explicitly center research as a political project allied in struggles against oppression and domination, and (2) they argue for embedded research, seeking to break down the activist/academic division. This is not to say that these are the only such perspectives doing so; however, I turn to these perspectives for their emphasis and their explicitly activist and action-oriented commitments.
3. There are certainly more broad definitions; however, I find Hale's definition attractive for its specific orientation to activism.
4. This reflexive focus draws from and is located in anti-racist (Dei & Johal, 2005) and feminist methods (Mohanty, 2003).
5. While Dei and Kempf's work provides a useful framework for anti-colonialism, there is a tendency in defining colonialism as anything that might be dominating (Dei, 2006, p. 3) that empties out its historical specificity. Such a broad definition loses the contexts in which colonialism was erected and also its specific impacts. If anything colonialism needs to be appropriately contextualized clarifies the specificities of oppression, domination and resistance that have occurred. This is a point Kempf (2010, p. 18) acknowledges, while contradictorily holding to the broad understanding above—a contradiction which cannot be further elaborated here. Ultimately, what is needed is a historically specific understanding of colonialism alongside an intersectional understanding of oppression.

References

Absolon, K. & Willet, C. (2005) Putting ourselves forward: Location in aboriginal research, in: L. Brown & S. Strega (Eds) *Research as Resistance: Critical, Indigenous and Anti-Oppressive Approaches*, pp. 97–126 (Toronto: Canadian Scholars' Press/Women's Press).

Alfred, T. & Corntassel, J. (2005) Being indigenous: Resurgences against contemporary colonialism, *Government and Opposition*, 40(4), pp. 597–614.

Amster, R., Deleon, A., Fernandez, L. A., Nocella, A. J. & Shannon, D. (Eds) (2009) *Contemporary Anarchist Studies: An Introductory Anthology of Anarchy in the Academy* (New York: Routledge).

Barker, A. J. (2009) The contemporary reality of Canadian imperialism: Settler colonialism and the hybrid colonial state, *American Indian Quarterly*, 33(3), pp. 325–351.

Barker, A. (2010) From adversaries to allies: Forging respectful alliances between indigenous and settler peoples, in: L. Davis (Ed.) *Alliances: Re/Envisioning Indigenous-non-Indigenous Relationships*, pp. 316–333 (Toronto: U of T Press).

Battiste, M. (Ed.) (2002) *Reclaiming Indigenous Voice and Vision* (Vancouver: UBC Press).

Cannella, G. S. & Manuelito, K. D. (2008) Feminisms from unthought locations: Indigenous worldviews, marginalized feminisms, and revisioning an anticolonial social science, in: N. K. Denzin, Y. S. Lincoln & L. T. Smith (Eds) *Handbook of Critical and Indigenous Methodologies*, pp. 45–60 (Thousand Oaks, CA: Sage).

Casas-Cortés, M. & Cobarrubias, S. (2007) Drifting through the knowledge machine, in: D. Graeber & S. Shukaitis (Eds) *Constituent Imagination: Militant Investigations/Collective Theorization*, pp. 112–126 (Oakland, CA: AK Press).

Cox, L. & Barker, C. (2002) What have the Romans ever done for us? Academic and activist forms of movement theorizing, alternative futures and popular protest, Paper presented at the 8th Annual Conference, Manchester Metropolitan University, April 2002. Available at http://www.iol.ie/~mazzoldi/toolsforcha nge/afpp/afpp8.html (accessed 4 January 2011).

Dei, G. J. S. (2006) Introduction: Mapping the terrain—Towards a new politics of resistance, in: G. J. S. Dei & A. Kempf (Eds) *Anti-Colonialism and Education: The Politics of Resistance*, pp. 1–23 (Rotterdam, The Netherlands: Sense Publishers).

Dei, G. J. S. (2010) Afterward: The anti-colonial theory and the question of survival and responsibility, in: A. Kempf (Ed.) *Breaching the Colonial Contract: Anti-Colonialism in the U.S. and Canada*, pp. 251–257 (Breinigsville, PA: Springer Science + Business Media).

Dei, G. J. S. & Asgharzadeh, A. (2001) The power of social theory: The anti-colonial discursive framework, *Journal of Educational Thought*, 35(3), pp. 297–323.

Dei, G. J. S. & Johal, G. S. (Eds) (2005) *Critical Issues in Anti-Racist Research Methodologies* (New York: Peter Lang).

Denzin, N. K., Lincoln, Y. S. & Smith, L. T. (Eds) (2008) *Handbook of Critical and Indigenous Methodologies* (Thousand Oaks, CA: Sage).

Doxtater, M. G. (2004) Indigenous knowledge in the decolonial era, *American Indian Quarterly*, 28(3–4), pp. 618–633.

Fernandez, L. A. (2009) Being there: Thoughts on anarchism and participatory observation, in: R. Amster, A. Deleon, L. A. Fernandez, A. J. Nocella & D. Shannon (Eds) *Contemporary Anarchist Studies*, pp. 93–102 (New York: Routledge).

Fischman, G. E. & McLaren, P. (2005) Rethinking critical pedagogy and the Gramscian and Freirean legacies: From organic to committed intellectuals or critical pedagogy, commitment, and praxis, *Cultural Studies* ↔ *Critical Methodologies*, 5(4), pp. 425–447.

Frampton, C., Kinsman, G., Thompson, A. K. & Tilleczek, K. (Eds) (2006) *Sociology for Changing the World: Social Movements/Social Research* (Winnipeg: Fernwood).

Getty, G. A. (2010) The journey between western and indigenous research paradigms, *Journal of Transcultural Nursing*, 21(1), pp. 5–14.

Gordon, U. (2007) Practicing anarchist theory: Towards a participatory political philosophy, in: D. Graeber & S. Shukaitis (Eds) *Constituent Imagination: Militant Investigations/Collective Theorization*, pp. 276–287 (Oakland, CA: AK Press).

Graeber, D. & Shukaitis, S. (Eds) (2007) *Constituent Imagination: Militant Investigations/Collective Theorization* (Oakland, CA: AK Press).

Grande, S. (2008) Red pedagogy: The un-methodology, in: N. K. Denzin, Y. S. Lincoln & L. T. Smith (Eds) *Handbook of Critical and Indigenous Methodologies*, pp. 233–254 (Thousand Oaks, CA: Sage).

Hale, C. R. (2006) Activist research v. cultural critique: Indigenous land rights and the contradictions of politically engaged anthropology, *Cultural Anthropology*, 21(1), pp. 96–120.

Hales, J. (2006) An anti-colonial critique of research methodology, in: G. J. S. Dei & A. Kempf (Eds) *Anti-Colonialism and Education: The Politics of Resistance*, pp. 243–256 (Rotterdam, The Netherlands: Sense Publishers).

Haluza-DeLay, R. (2003) Community-based research, movement intellectuals and the 'knowledge council', *Canadian Review of Social Policy*, 52, pp. 133–138.

Jones, A. with Jenkins, K. (2008) Rethinking collaboration: Working the indigene-colonizer hyphen, in: N. K. Denzin, Y. S. Lincoln & L. T. Smith (Eds) *Handbook of Critical and Indigenous Methodologies*, pp. 471–486 (Thousand Oaks, CA: Sage).

Juris, J. S. (2007) Practicing militant ethnography with the movement for global resistance in Barcelona, in: D. Graeber & S. Shukaitis (Eds) *Constituent Imagination: Militant Investigations/Collective Theorization*, pp. 164–178 (Oakland, CA: AK Press).

Kempf, A. (2010a) Introduction: The politics of the North American colonial in 2009, in: A. Kempf (Ed.) *Breaching the Colonial Contract: Anti-Colonialism in the U.S. and Canada*, pp. 1–34 (Breinigsville, PA: Springer Science + Business Media).

Kempf, A. (2010b) Contemporary anti-colonialism: A transhistorical perspective, in: A. Kempf (Ed.) *Breaching the Colonial Contract: Anti-Colonialism in the U.S. and Canada*, pp. 13–34 (Breinigsville, PA: Springer Science + Business Media).

Kempf, A. (2006) Anti-colonial historiography: Interrogating colonial education, in: G. J. S. Dei & A. Kempf (Eds) *Anti-Colonialism and Education: The Politics of Resistance*, pp. 129–158 (Rotterdam, The Netherlands: Sense Publishers).

Kinsman, G. (2006) Mapping social relations of struggle: Activism, ethnography, social organization, in: C. Frampton, G. Kinsman, A. K. Thompson & K. Tilleczek (Eds) *Sociology for Changing the World: Social Movements/Social Research*, pp. 133–156 (Winnipeg: Fernwood).

Kovach, M. (2005) Emerging from the margins: Indigenous methodologies, in: L. Brown & S. Strega (Eds) *Research as Resistance: Critical, Indigenous and Anti-Oppressive Approaches*, pp. 19–36 (Toronto: Canadian Scholars' Press/Women's Press).

Kovach, M. (2009) *Indigenous Methodologies: Characteristics, Conversations and Contexts* (Toronto: University of Toronto Press).

Laenui, P. (Hayden F. Burgess) (2002) Processes of decolonization, in: M. Battiste (Ed.) *Reclaiming Indigenous Voice and Vision*, pp. 150–160 (Vancouver: UBC Press).

Latif, A. & Jeppesen, S. (2007) Toward an anti-authoritarian anti-racist pedagogy, in: D. Graeber & S. Shukaitis (Eds) *Constituent Imagination: Militant Investigations/Collective Theorization*, pp. 288–300 (Oakland, CA: AK Press).

Little Bear, L. (2002) Jagged worldviews colliding, in: M. Battiste (Ed.) *Reclaiming Indigenous Voice and Vision*, pp. 77–85 (Vancouver: UBC Press).

Louis, R. P. (2007) Can you hear us now? Voices from the margin: Using indigenous methodologies in geographic research, *Geographical Research*, 45(2), pp. 130–139.

Max, K. (2005) Anti-colonial research: Working as an ally with aboriginal peoples, in: G. J. S. Dei & G. S. Johal (Eds) *Critical Issues in Anti-Racist Research Methodologies*, pp. 79–94 (New York: Peter Lang).

McClintock, A. (1992) The angel of progress: Pitfalls of the term 'post-colonialism', *Social Text*, 31–32, pp. 84–98.

Mohanty, C. T. (2003) *Feminism Without Borders: Decolonizing Theory, Practicing Solidarity* (Durham: Duke University Press).

Regan, P. (2010) *Unsettling the Settler Within: Indian Residential Schools, Truth Telling and Reconciliation in Canada* (Vancouver: UBC Press).

Routledge, P. (2009) Toward a relational ethics of struggle: Embodiment affinity, and affect, in: R. Amster, A. Deleon, L. A. Fernandez, A. J. Nocella & D. Shannon (Eds) *Contemporary Anarchist Studies*, pp. 82–92 (New York: Routledge).

Simpson, L. R. (2004) Anticolonial strategies for the recovery and maintenance of indigenous knowledge, *American Indian Quarterly*, 28(3–4), pp. 373–384.

Smith, G. H. (2002) Protecting and respecting indigenous knowledge, in: M. Battiste (Ed.) *Reclaiming Indigenous Voice and Vision*, pp. 209–224 (Vancouver: UBC Press).

Smith, L. T. (2004) *Decolonizing Methodologies: Research and Indigenous Peoples* (London: Zed Books/ Dunedin: University of Otago Press).

Thomas, R. A. (2005) Honouring the oral traditions of my ancestors through story telling, in: L. Brown & S. Strega (Eds) *Research as Resistance: Critical, Indigenous and Anti-Oppressive Approaches*, pp. 237–254 (Toronto: Canadian Scholars' Press/Women's Press).

Wilson, S. (2008) *Research is Ceremony: Indigenous Research Methods* (Winnipeg: Fernwood).

Disclosed and Willing: Towards A Queer Public Sociology

ANA CRISTINA SANTOS

Centre for Social Studies, University of Coimbra, Coimbra, Portugal

ABSTRACT   *This article contributes to recent debates on 'public sociology', expanding the notion and interrogating its utility for those who simultaneously carry out activism and scholarship. The idea of public sociology has underpinned the conviction that knowledge can contribute to inclusion or exclusion, depending on how it is used. This article argues that commitment to public sociology implies abiding by the guiding principles of accountability, intersectionality, reciprocity and reflexivity, and further represents commitment to activism, embracing politics as an intended effect of knowledge production. Building on personal experience as researcher and activist in the lesbian, gay, bisexual and transgender movement in Portugal, I also explore the epistemological and ethical impacts of taking on the role of scholar-activist. This offers a 'double agency' through which one may build and disseminate empirically grounded knowledge whilst maintaining a sense of social responsibility and political engagement. Bringing these ideas together, this article advances the notion of a 'queer public sociology': a critical framework that accounts for sexual diversity, and that acknowledges its politically situated character at the same time that it contributes to the dismantling of sexual prejudice and exclusion.*

1. Introduction

Despite being highly contested, the legacy of positivism in sociological thought is still pervasive today. This legacy is mirrored by the ways in which sociology frequently operates according to dominant ways of thinking and doing, rather than being proactively engaged in tackling inequality. The notion of public sociology, initially advanced by Herbet J. Gans (2002), was crucial in moving away from positivist approaches within mainstream sociology. Drawing on the notion of public sociology, and inspired by feminist and queer perspectives on knowledge production and the research process, this article considers the importance of disclosing the inevitable political engagement of sociological work to render it more plausible, accountable and, ultimately, useful.

In the first part of the article, I expand on the notion of public sociology (Gans, 2002; Burawoy, 2004a, 2004b, 2004c, 2005). Underpinning the idea of public sociology is the conviction that knowledge can contribute to processes of inclusion or exclusion, depending on how it is used. As feminist methodologies also suggest, the ultimate purpose of knowledge production should be to reach audiences who are not necessarily related to academia (Harding, 1991, 2004; Haraway, 2004).

Considering my personal experience as a researcher and activist in the lesbian, gay, bisexual and transgender (LGBT) movement in Portugal, in the second part of the article, I explore the epistemological and ethical impacts of being a disclosed activist and academic in LGBT and queer issues. I argue that this type of 'double agency' offers the opportunity to build and disseminate empirically grounded knowledge whilst maintaining a sense of social responsibility and political engagement.

Finally, against a positivist understanding of science, in the last part of this article, I advance the notion of a 'queer public sociology' (QPS), i.e. a critical framework that accounts for sexual diversity, and that acknowledges its politically situated character at the same time that it contributes to the dismantling of sexual prejudice and exclusion. I sustain that it is time to interact politically with a world whose realities of social exclusion and inequality demand a proactive role from academics, particularly in the intersecting field of sociology and LGBT and queer studies.

2. Public Sociology in Social Movement Studies[1]

Alain Touraine's sociology of action suggested that the researcher should become what might be interpreted as a Gramscian hybrid of the traditional intellectual and the organic intellectual (Gramsci, 1971). This would be the role of the intellectual who, without abandoning their ivory tower, aims also to solve the hermeneutic and communication gaps between actor and opponents, promoting what Touraine labels 'permanent sociology' (1981, p. 148), which would cast light upon problems deriving from collective action. When one reads these early writings of Touraine, there is an almost inescapable sense of a scholar who is, albeit unwillingly, patronising social movements and activists, as if science, or indeed scientists themselves, were (necessarily) particularly enlightened. Such an approach is indeed hard to sustain when one recognises that all knowledge is situated (Haraway, 2004; Harding, 2004), including that which is produced in academic contexts.

The sort of privileged role that Touraine ascribes to academic knowledge has been under siege for a number of years, particularly by feminist, LGBT and queer authors who argue that the interventions of scholars also contain the potential for distortion, bias and error, inasmuch as other forms of knowledge do (Harding, 1991, 2004; Haraway, 2004; Ahmed, 2006). In 1987, Frigga Haug co-edited a book on the uses of memory work as a method that could counter the shortcomings of non-agentic positivist science. She then suggested that we searched 'for possible indications of how we have participated actively in the formation of our own past experience' (1987, p. 35), as a way to abandon what she perceives as 'the usual mode of social-scientific research, in which individuals figure exclusively as objects' (1987, p. 35). Accordingly, by generating empirically grounded knowledge, memory work was, as Anne-Jorunn Berg phrased it, a 'suitable method to help bridging the gap between social theory and experience' (2008, p. 215). Memory work, parallel to autobiographic and other narrative methods, was introduced by feminist scholars who felt discouraged by the excluding *modus operandi* advanced by positivist models (Oakley, 1982; Stanley, 1991).

The desire for more permanent ways of bringing academia closer to everyday experience has also inspired the notion of public sociology. As initially advanced by Gans, public sociology is an alternative to the notion of the public intellectual:

A public sociologist is a public intellectual who applies sociological ideas and findings to social (defined broadly) issues about which sociology (also defined broadly) has something to say. Public intellectuals comment on whatever issues show up on the public agenda; public sociologists do so only on issues to which they can apply their sociological insights and findings. They are specialist public intellectuals. (2002, p. 2)

Public sociology was originally presented as a theoretical approach that acknowledged the highly contingent framework of scientific production as well as science's responsibility in liaising with other actors to develop reciprocal and non-hierarchic learning processes. Drawing on Gans' work, Michael Burawoy suggested that:

The bulk of public sociology is indeed of an organic kind—sociologists working with a labor movement, neighborhood associations, communities of faith, immigrant rights groups, human rights organizations. Between the organic public sociologist and a public is a dialogue, a process of mutual education. The recognition of public sociology must extend to the organic kind which often remains invisible, private, and is often considered to be apart from our professional lives. The project of such public sociologies is to make visible the invisible, to make the private public, to validate these organic connections as part of our sociological life. (2005, pp. 7–8)

Several aspects in this excerpt deserve further commentary. First, Burawoy's definition of public sociology seems to imply a bilateral (or even multifarious) process of exchange, 'a dialogue' that aims at enhancing reciprocal chances of learning. Second, such process involves academia, but also the wider society ('a public') that is expected to be recognised by sociologists as equally important interlocutors in this dialogue. Third, Burawoy's arguments contain an implicit call for politicised action: sociologists have the power, and the duty, to intervene in the social sphere to enhance visibility, participation and inclusion. As such, political engagement is not merely an unintended consequence of sociological work; it is rather a process of willing disclosure through which sociologists become engaged political actors. Furthermore, such engagement is clearly influenced by feminist writings and demands that have been ground-breaking in advancing the notion that the personal is political, and the private should be public (Oakley, 1982; Lister, 1997; Harding & Norberg, 2005; Ryan-Flood & Gill, 2010). Finally, public sociology is not a mere 'add-on', something external to the sociological work itself, but a vital part of it.

Such politicised understanding of sociology is sustained on several occasions in Burawoy's work (2004b, 2005). According to him, sociologists

constitute an actor in civil society and as such have a right and an obligation to participate in politics. [. . .] The 'pure science' position that research must be completely insulated from politics is untenable since antipolitics is no less political than public engagement. (2004b, p. 1605)

In other words, it is time sociologists interact politically with a world whose realities of exclusion and inequality demand a proactive role from academics and from sociologists in particular. In accordance with this rationale, knowledge production should be concerned with audiences beyond academia, investing in outreaching initiatives that disseminate research findings in an accessible language and engaging different types of social actors during the process of knowledge production (Ackerly & True, 2010; Taylor & Addison, 2011). One example may better illustrate this. Writing in 2004, Charlotte Ryan described her successful joint experience with anti-racist organisations regarding local TV news stations in Boston. There were concerns about crime reports reinforcing racist perceptions. Sociologists and activists worked together, campaigning for news coverage to put crime in economic and political contexts. According to Ryan, 'it represented public sociology at its best, synergistically linking uncommon partners to deepen knowledge and equalize social resources' (2004, p. 112). This example highlights how grounded theory can be a crucial sociological tool 'to prove that there are other things to be known through other ways of knowing' (Widerberg, 2008, p. 113). Perhaps more importantly, such intersection between academia and civil society is the condition to achieve social and cognitive justice (de Sousa Santos, 2006). It also starts to pave the way for the inclusion of intersectionality[2] as a fundamental aspect of politically engaged research, particularly in the field of feminist and LGBT and queer studies (Valentine, 2007; Cole, 2008; Davis, 2008; Shields, 2008; Taylor *et al.*, 2010).

Social movement studies offer sociologists the opportunity to strengthen mutual intelligibilities between academics and activists. Indeed, sociology emerged from the need to understand how societies operate and change, and how people respond to—*intervene* in—that change. It was the transformation introduced after the industrial revolution that prompted scholars such as Auguste Comte, Harriet Martineau, Henri de Saint Simon, Max Weber and Emile Durkheim, to name a few, to start a new discipline that regarded change as a social fact. Without this symbiotic element of change and intervention, sociological inquiry would be fundamentally voided and there would not be much left to be discussed in or examined by sociology or even, more generally, by the social sciences. Burawoy has referred to this by alerting that 'the professional temptation toward insularity and abstraction threatens to cut off sociology's lifeblood that comes from connection to the concrete world beyond (2004c, p. 105). William Gamson has phrased this necessary link along similar lines: 'public sociology has helped to keep my professional sociology grounded in the real world' (2004, p. 107). These lines of thought are clearly inspired by feminist epistemology and ethics regarding the research process (Haraway, 2004; Harding, 2004; Harding & Norberg, 2005; Ackerly & True, 2010; Ryan-Flood & Gill, 2010).

Given the above observations, it seems rather obvious that the social sciences in general, and sociology in particular, are historically depended on elements such as participation and change, rather than being intrinsically connected to processes taking place exclusively within academic institutions. Despite this evidence, professional sociology remains largely wary of compromise, co-production and interdependence, which has impacted the way sociological theory is used by activists. Studying the topic of relevance of social movement studies, Dick Flacks's findings are of interest to my argument in this article, particularly when he concludes that activists do not engage with the existing literature on the sociology of social movements, opting instead for reading history, biographies and memoirs (2005, p. 59). If this is indeed so, the relevance of current studies of social movements is undoubtedly compromised. Rather than dismissing such findings with a quick shrug, perhaps it is more useful to address the interpretative gap between the cognitive horizons represented in academic and activist discourses, and try to counter it. In fact, recognising this lack of mutual intelligibility may constitute a first step towards new, less closeted forms of knowledge, more widely available, informing bottom-up dialogues and enabling reciprocal learning processes between academics and advocates. Ultimately, it may even persuade activists to pursue their studies and academics to engage in militancy. Flacks examines the advantages associated with investing in joint work gathering activists and researchers. According to him, this would strategically enhance the scope and efficiency of the knowledge available to both parties (2005, p. 54). And, indeed, it might lead to what Bevington & Dixon (2005) call a 'movement-relevant theory'.

A similar approach has been suggested by Boaventura de Sousa Santos, who advanced the notion of 'ecology of knowledges' as a way to foster mutual intelligibility and cooperation between academic and non-academic institutions and people. According to him, this 'advanced form of action-research' (2006, p. 78) represents an epistemologically revolutionary departure from the conventional ways of knowledge production. Drawing on critical theory approaches, de Sousa Santos continues to argue that mainstream academic knowledge has often disregarded a vast array of sources and interlocutors, causing 'the impoverishment of human experience and diversity' (2006, p. 81) as well as compromising both social and cognitive justice.

A commitment to public sociology necessarily represents a shift in the research ethics underpinning epistemological and methodological choices to the extent that contributing to social justice becomes a central aim of knowledge production. This is particularly relevant when the topic of research has historically been subject to discrimination and inequality, as it is the case with LGBT and queer issues. Methodologically, this would imply favouring plural data generation methods and analytical techniques—triangulation—as a way to benefit from different perspectives and analytical insights, rather than making knowledge production dependent on single, top-down contributions. From an ethical point of view, a commitment to public sociology implies abiding by the guiding principles of accountability, intersectionality, reciprocity and reflexivity, which will be detailed later in the article. It would also represent a commitment to activism as a significant part of citizenship and an embracement of politics as an intended effect of knowledge production. It would, ultimately, lead to a willing disclosure of the political engagements of scholar-activists, i.e. those who are simultaneously academics and activists. The implications of such disclosure will be discussed in the next section.

3. Disclosed LGBT Activism Within Academia

Activism can be defined as a voluntary engagement in struggles for recognition and/or redistribution (Fraser & Honneth, 2003). It is always a political act because it implies a public commitment to a cause. As such, activism must be about public participation, even when this intervention is virtual and mostly mediated by electronic devices (mobile phones or social networks, for instance). This form of politicised intervention often takes the shape of collective action in organisations or social movements. However, it can also consist of sporadic mobilisation for particular purposes, in particular contexts, such as protest (Della Porta & Diani, 1999; Goodwin & Jasper, 2009) or simply 'everyday acts of defiance' (Baumgardner & Richards, 2000, p. 283). Regardless of its more or less regular character, activism is about citizenship, in the sense that it draws on the right to intervene and to be recognised. The notion of activism, as explicated above, does not seem to be incompatible with scholarly production in the realm of sociology (or any other, for that matter), precisely because social intervention sits at the core of sociological inquiry.

If this is indeed the case, should we, as researchers, sacrifice acknowledging our political standpoints for the positivist sake of retaining an allegedly value-neutral objectivity, which is after all a 'weak objectivity' (Harding, 1991)?[3] Arguably, sociology benefits from disclosed political engagements, to the extent that sociologists are, themselves, actors in processes and facts under sociological scrutiny. What seems artificial, then, is the alleged distinction between science and politics, as if a strict boundary, however fake and precarious, could secure scientific accuracy. I suggest that what is wrong in this equation is the premise of neutrality, which disregards the fundamental fact that all actors, including sociologists, are situated subjects.

To the extent that context informs people's standpoints—from which, then, sociology is produced—it is not possible to escape a knowledge that is inextricably bounded and situated. Then, the next logic step, it seems, would be to recognise one's political standpoint and to strive for a 'strong objectivity', defined by Harding as 'a commitment to acknowledge the historical character of every belief or set of beliefs' (1991, p. 156). Harding underlines the inescapability of 'historical gravity' by saying that

> Political and social interests are not 'add-ons' to an otherwise transcendental science that is inherently indifferent to human society; scientific beliefs, practices, institutions, histories, and problematics are constituted in and through contemporary political and social projects, and always have been. (1991, p. 145)

Speaking as a standpoint theorist and arguing against the 'conventional view ... [that] politics can only obstruct and damage the production of scientific knowledge' (2004, p. 1), she correctly points out that

> The more value-neutral a conceptual framework appears, the more likely it is to advance the hegemonous interests of dominant groups, and the less likely it is to be able to detect important actualities of social relations [...]. The 'moment of critical insight' is one that comes only through political struggle. (2004, pp. 6, 9)

Wylie takes the argument of the usefulness of political engagement a step further, writing that 'considerable epistemic advantage may accrue to those who approach inquiry from an interested standpoint, even a standpoint of political engagement' (2004, p. 345). Though an extended debate about standpoint theory and its critiques is beyond the scope of this article, I want to emphasise the importance of political engagement within academia.

As Harding eloquently put it, standpoints are 'toolboxes enabling new perspectives and new ways of seeing the world to enlarge the horizons of our explanations, understandings and yearnings for a better life' (2004, p. 5). In this context, 'double-agency'—understood as the politically engaged role of scholar-activists within academia—becomes not only legitimate, but desirable. The possibility of a desirable role for scholar-activists within academia is clearly informed by the notion of public sociology, as discussed in the previous section.

The acknowledgement of interdependence and the call for intersectionality between academia and civil society represent a new ethics of research, committed to the willing disclosure of researchers' political engagement. In the field of LGBT and queer studies, such engagement is politicised to the extent that the choice of topic is already political. In a context in which discrimination represents invisibility, oppression and violence, sociologists who study LGBT or queer issues are certainly expected to use relevant knowledge and resources to counter the effects of such discrimination (Santos, 2006a, 2008). Discrimination heightens the call for sociologists to become scholar-activists. As noted by Halberstam, 'The academic might be the archivist or a co-archivist or they might be a fully-fledged participant in the subcultural scene that they write about. Only rarely does the queer theorist stand wholly apart from the subculture, examining it with an expert's gaze' (2003, p. 322). Therefore, disclosing political engagement within academia becomes not only a possibility but also a duty in relation to the dominant framework of sexism, heterosexism and homo-, bi- or transphobia.

In his book, *The Unfinished Revolution*, Engel offers an example of a study situated at the junction between academia and activism. Engel states that his participation in Washington's candlelight vigil for the murder of the young gay man Matthew Shepard in October 1998 made him ascribe a new meaning to his research, as he realised that 'an emotionally emptied account of this movement fails to do justice to the individuals who work every day so that gays, lesbians, bisexuals and transgender people can live safer and happier lives' (2001, p. 3). This event impelled Engel to write a book with a pragmatic goal: hoping that the evolution of social theory on social movements would allow for a deeper understanding of gay and lesbian movements. He believed that ultimately such a task could help LGBT movements to learn how to benefit from political opportunities, so that homophobia and heterosexism would finally be overturned. Engel's stated purpose of the usefulness of his research reveals the potential for engagement between academia and activism. Furthermore, it highlights that, rather than seeking to minimise one's impact, one should self-reflexively acknowledge it.

Another example of LGBT activism within academia is currently presented by most Portuguese scholars who do research on LGBT and queer issues. Despite being a relatively recent field in the country, there is an interesting overlap between the roles of activist and academic, with the majority of published material being written by authors who are publicly known to be both academics and LGBT and queer activists (Cascais, 2004, 2006; Santos, 2005, 2006b, 2008; Carneiro & Menezes, 2007; Almeida, 2009; Carneiro, 2009; Oliveira *et al.*, 2009; Nogueira & Oliveira, 2010). Such overlap can be partially explained by the story of LGBT activism in Portugal. This story is almost as recent as the academic field of LGBT and queer studies, and it dates back to the 1990s. After 48 years of dictatorship and in a country highly influenced by Catholic morals, the movement presented a late, but steady, development. The country's legislation on LGBT rights is amongst the most advanced in Europe, including a *de facto* unions law (since 2001), an anti-discrimination constitutional provision (since 2004), equal ages of consent (since 2007), protection against same-sex domestic violence hate crimes (since 2007), civil marriage (since 2010) and a gender recognition provision (since 2011). These changes were to a great extent accelerated by the LGBT movement—including engaged academics—and influenced by wider processes of modernisation and Europeanisation. Throughout the 2000s, the movement was able to make use of supportive journalists, politicians and academics, fostering the sort of engagement that granted political legitimacy and social acceptability to the movement's demands (Cascais, 2006; Carneiro & Menezes, 2007; Santos, 2008).

LGBT political engagement of academics impacts upon epistemological and ethical choices because struggling against discrimination becomes a permanent concern, a personal commitment and a fundamental aim in any research process. Such engagement has also contributed to community-building and mutual academic support, which is particularly relevant in a country where LGBT and queer studies still face hostility and scepticism within mainstream academia. A personal story may add to this argument. I entered the realm of LGBT studies via academia, when I was preparing a thesis for a degree in Sociology in the 1990s. By the time I had finished writing the thesis, I was already a regular participant in LGBT events and a few years later I co-founded a queer youth organisation. Since then, participating in LGBT collective action has been as important as other academic commitments. There were uncountable personal and professional gains resulting from this double agency, and I have always tried to focus on those to counter the occasional setback along the way.[4] Many of my colleagues share the same experience. There is indeed empirical evidence of the common perception that 'we need not—indeed, must not—choose between "good politics" and "good science" [. . .] for the former can produce the latter', as Harding suggests (2004, p. 6).

A central point of this article is that our multiple belongings impact on our knowledge production in ways that are always political, whether we acknowledge it or not. As Stephen Pfohl puts it, 'our own personal and institutional locations within matrices of power always partially shape what we see and what escapes our sight. [. . .] We are never simply ourselves alone, but always also complex social personae, enacting cultural scripts not entirely of our own making' (2004, pp. 114–115). The previous examples highlight the multiple, and often conflictual, belongings of social actors.

By revaluing the notion of standpoint, rather than attempting to shield science from politics, scholar-activists are contributing to a significant sociological turn, one that reinvents sociology as a socially and politically relevant field of studies. This turn presents opportunities, as well as challenges, stemming from the epistemological and ethical implications of political engagement.

Scholar-activists are in a privileged position to access target groups—including policy and law-makers, politicians and the media—that can be crucial agents for enacting social, legal and political change. More specifically in the field of LGBT and queer studies, scholar-activists are invested with the trust and hope of social actors who experience the ongoing effects of daily discrimination based on sexual orientation and gender identity. This equips scholar-activists with the ethical duty of producing science that is accessible to a general audience and disseminated amongst groups and institutions that have the power to counter discrimination. Arguably, this willingness to disclose oneself as a politically engaged academic will make sociology more socially and scientifically accountable and, equally importantly, more relevant for countering inequality and discrimination, within academia and beyond.[5]

This sociological turn also presents challenges. Stephen Pfohl refers to these as a terrible lesson and a curse, 'the curse of no longer being able to easily exercise white, heterosexist, or class-based privilege without pangs of conscience' (2004, p. 115). Arguably, the sociological turn brought about by the disclosure of activism within academia highlights the need for a new critical framework within sociology. I shall return to this topic in the last section of this article.

4. Towards A Queer Public Sociology

Just as 'Sociology is born with civil society and dies with civil society' (Burawoy, 2004b, p. 1615), early lesbian and gay studies were inspired by the development of LGBT organisations, particularly in the USA. Such connection between activism and scholarship expanded to other parts of the globe (Altman, 1996). The emergence of LGBT movements, with visible and autonomous claims vis-à-vis the civil rights movement, provided content for the academic debates and theories of sexuality during the 1960s and the 1970s. In this regard, the events related to the Stonewall riots, for instance, gave a significant nudge to both academia and beyond.[6] The LGBT movement itself had been inspired by other theoretical developments. Emerging in the western world during the 1960s, lesbian and gay studies were largely informed by the theoretical perspective of new social movement studies, which focused mostly on exploring the reasons underlying collective action (Foweraker, 1995, p. 2).

This connection between academia and activism in the realm of LGBT and queer studies has been explored by Medhurst & Munt, who ask, 'Is there something called Lesbian and Gay Studies? There cannot be such a thing without the lesbian and gay identities and communities which inform them and are simultaneously constructed by them' (1997, p. xiii). Drawing on the early lesbian and gay studies, Altman agrees with Medhurst & Munt and provides a historical context for such intersection: 'Like related works in sociology and political science these works were firmly grounded in the movement, and the first generation of gay and lesbian scholars were also activists' (1996, p. 4). A similar process took place in the UK, where many academics in the field of LGBT and queer studies were also active members of LGBT organisations. That was the case with Jeffrey Weeks, Ken Plummer and Mary McIntosh, amongst others, who were engaged in Gay Liberation, at the same time as they were engaged in their respective careers in academia (Medhurst & Munt, 1997).

Therefore, LGBT and queer issues have always connected academia and activism from the outset, under the premise that such connection was essential to provide accountability and accuracy to scientific knowledge (Irvine, 2003). Arguably, one could say that activism is the reality-check of LGBT and queer studies.

Considering the specific case of queer studies, these stemmed from the alleged inefficiency of feminist theory and early lesbian and gay studies regarding the politicisation of sexual-related issues characterised by their uncertainty, fluidity and elasticity (Santos, 2006a, b; Giffney & O'Rourke, 2009). The increasing complexity of social facts and phenomena, linked to the emergence of new identities and groups rendered sexually dissident, pushed academia into new theoretical tools from a range of disciplines. A new field of studies was born. As Noreen Giffney has described it

> Queer has many centres. Just as no one discipline can lay claim to the pastiche that is queer theory, so too can no one theoretical discourse or mode of enquiry broadcast its ownership. (2004, p. 74)

Despite the interdisciplinary character of queer studies, the connection between sociology—particularly the perspective of public sociology—and queer studies remains scarce. According to Irvine, there is a reason behind the absent link between sociology and sexuality studies in general

> I would like to suggest one additional reason why sociology tends to marginalize sexuality studies: it is a stigmatized subject casting suspicion upon those who study it. The history of sexuality research throughout the twentieth century has been one of

stigma. The topic is controversial, even disreputable to many, and researchers have been repeatedly warned against studying sexuality. (Irvine, 2003, p. 451)

Nevertheless, some authors acknowledge the gains of promoting a sociological input in LGBT and queer studies. That is the case of Steven Seidman, according to whom

> Queer theory suggests to sociologists a more reflexive analysis of sexual categories and the ways these sexual meanings intersect with institutions to shape dynamics of order and oppression. Sociologists, in turn, have something crucial to offer: a rich tradition of social-structural and cultural analysis that can give empirical richness to the often literary or abstract conceptual analyses of Queer theory. (1996, p. 17)

After acknowledging both the sociological deficit of queer theory and the queer deficit of sociology, Ki Namaste suggested a queer sociological theory that would invest in the transformation of sexual politics to overcome the binaries that characterise dominant frames (1996, pp. 205–206).

In line with this strand of argumentation, I advance the notion of queer public sociology (QPS) to describe a new critical framework invested in changing public policy, law and political and cultural institutions as a way to tackle discrimination based on sexual orientation and gender identity. QPS is proactively engaged in action research,[7] methodological triangulation and ethical principles guided by the goals of accountability and reciprocity, intersectionality, reflexivity and sexual justice. According to this perspective, issues such as sexism, heterosexism, homo-, bi- and transphobia become a representational system, a social construction that demands to be studied to be overturned. Discrimination is a collective product that stems from unequal power relations, instead of an individual problem. Therefore, the focus is moved away from the (individual) victim to the structural system that enables and legitimises discrimination. QPS is inextricably linked to the duty to inform the struggle against such structural discrimination and, as such, it willingly embraces political engagement both as an epistemological and as an ethical choice.

The analysis of the mutual implications between sociology and LGBT and queer studies suggests a series of guiding principles regarding the ethics of queer public sociologies.

Accountability and *reciprocity* can be jointly considered as a first guideline, consisting of building rapport and respecting step-by-step agreements between the different participants in the research process. It also includes retribution of the input participants offer to the research outcome. This may imply service provision, voluntary work and sharing resources accumulated during the research process (e.g. media analysis, databases, annotated bibliographies, etc). An 'ethics of care' (Roseneil, 2004; Held, 2006) should therefore support the work of scholar-activists, particularly in situations that involve vulnerability and oppression, such as study and advocacy in the fields of same-sex sexuality and gender identity.

Second, *intersectionality* should be regarded as a fundamental resource of QPS, focusing on the explanatory potential of a range of identities, contexts and locations. An intersectional approach grants sociological inquiry transversal and interdisciplinary analytical tools that offer greater accuracy and validity to the research process as a whole.

Third, *self-reflexivity* understood as a vigilant and systematic exercise of self-critique that stresses the researcher's responsibility and ethical compromise to reject sexually biased—homophobic, transphobic, biphobic, sexist and/or heterosexist—projects. The principle of self-reflexivity must also imply an interest in meta-theory, theory and methodological triangulation as a way to advance a 'power-reflexive methodology' (Pfohl, 2004, p. 115). Despite the critique of reflexivity as a profoundly classed product of late modernity (Skeggs, 2004) as well as 'potentially self-indulgent', harmful and partial (Sanchez-Taylor & O'Connell Davidson, 2010), this principle can perhaps be more usefully understood as a practice, a 'relationship [. . .] between being a sociologist and being a person' (Gouldner, 2004, p. 383). Therefore, self-reflexivity as it is being proposed in this article is about *doing*, rather than *being*.

Finally, as emphasised throughout the article, a commitment to *sexual justice* implies political engagement beyond the walls of academia, an epistemological and ethical decision to counter the boundaries of otherness that imply a gap between so-called 'science' and activism. If the goal is sexual justice, then activism becomes a duty of the responsible academic under the critical framework of QPS. As such, social and cognitive justices become inextricably connected.

These principles may be interpreted as a minimum conceptual standard for studies under the critical framework of QPS. They are not mutually excluding nor overriding, and they should certainly be challenged and adjusted to the empirically based needs of each particular study or project. However dynamic this critical framework should remain, the most innovative aspect of QPS is the ability to reject claims of scientific political disengagement, replacing them with the legitimacy of willing disclosure as a non-negotiable ethical choice.

5. Conclusion

This article set out to explore the importance of an increasing articulation between science and activism, and the impacts of such political engagement regarding epistemological decisions and the ethics of research. It argued that political engagement should be not only embraced but also publicly encouraged and celebrated as a way to improve the quality of sociological input, as well as the accountability and relevance of sociological findings.

Writing in early 20th century, Emile Durkheim affirmed

> The ideal society is not outside of the real society; it is part of it. [. . .] We cannot hold to one, without holding to the other. For a society is not made up merely of the mass of individuals who compose it, the ground which they occupy, the things which they use and the movements which they perform, but above all is the idea which it forms of itself. (1912, p. 189)

Likewise, the double ability to intervene as social actors and as sociologists should not be regarded as two poles that mutually repel each other, but rather as a necessary, dynamic and rewarding intersection.

To reiterate, academic production—whether research, lectures, presentations or publications—is always a result of situated knowledge. Rather than being almost embarrassed or trying to mask political engagement with the appearance of (always unattainable) neutrality, it is suggested that sociological theory will benefit from the multiple ways in which academia and politics intersect.

Disclosing one's double agency as scholar-activists is the necessary step to make knowledge production more relevant, as well as more transparent in its purposes and procedures. This seems to be particularly significant in the realm of LGBT and queer studies, in which issues of exclusion, discrimination and violence demand a sharp and informed intervention towards an inclusive future for all. QPS, as suggested in this article, is in a privileged position to contribute in a significant way to such important intervention.

Acknowledgements

I thank the Portuguese Foundation for Science and Technology for funding the research project 'Sexual Citizenship of Lesbian Women in Portugal' (reference PIHM/GC/0005/2008), from which this article results. I also thank comments from colleagues from the Centre for Social Studies and the Birkbeck Institute for Social Research to earlier versions of this article, as well as the insightful and encouraging remarks offered by Kevin Gillan, Jenny Pickerill and three anonymous reviewers of *Social Movement Studies*.

Notes

1. This section draws partially on Santos, 2006a.
2. Intersectionality can be understood as 'a tool for analysis, advocacy and policy development that addresses multiple discriminations and helps us understand how different sets of identities impact on access to rights and opportunities. [. . .] Intersectional analysis aims to reveal multiple identities, exposing the different types of discrimination and disadvantage that occur as a consequence of the combination of identities' (AWID, 2004, pp. 1–2).
3. According to Harding, 'Weak objectivity is located in a conceptual interdependency that includes (weak) subjectivity and judgemental relativism' (1991, p. 156).
4. For more on the topic of advantages and setbacks of being a scholar-activist in the Portuguese academia, see Santos, 2011.
5. Examples of LGBT and queer scholars coming out and making claims in and beyond academia include Davina Cooper, Zowie Davy, Ken Plummer, Diane Richardson, Sasha Roseneil and Jeffrey Weeks, amongst many others.
6. More information about the Stonewall events is available at http://www.stonewall.org.uk/ (accessed 03/01/2011).
7. For details on the advantages of bringing queer theory and action research together, see Filax, 2006.

References

Ackerly, B. & True, J. (2010) *Doing Feminist Research in Political and Social Science* (Basingstoke: Palgrave).
Ahmed, S. (2006) *Queer Phenomenology. Orientations, Objects, Others* (Durham, NC: Duke University Press).
Almeida, M. V. (2009) *A Chave do Armário. Homossexualidade, Casamento, Família* (Lisbon: Imprensa de Ciências Sociais).
Altman, D. (1996) On global queering, *Australian Humanities Review*, 2, pp. 1–7. Available at http://www.lib.la trobe.edu.au/AHR/archive/Issue-July-1996/altman.html (accessed 17 December 2004).
AWID – Association for Women's Rights in Development (2004) Intersectionality: a tool for gender and economic justice, *Women's Rights and Economic Change*, 9, pp. 1–8.
Baumgardner, J. & Richards, A. (2000) *Manifesta: Young Women, Feminism, and the Future* (New York: Farrar, Strauss & Giroux).
Berg, A. (2008) Silence and articulation: Whiteness, racialisation and feminist memory work, *NORA: Nordic Journal of Women's Studies*, 16(4), pp. 213–227.
Bevington, D. & Dixon, C. (2005) Movement-relevant theory: Rethinking social movement scholarship and activism, *Social Movement Studies*, 4(3), pp. 185–208.
Burawoy, M. (2004a) Public sociologies: Response to Hausknecht. Available at http://sociology.berkeley.edu/fa culty/burawoy/burawoy_pdf/PS.Hausknecht.vs.Burawoy.pdf (accessed 13 July 2006).
Burawoy, M. (2004b) Public sociologies: Contradictions, dilemmas, and possibilities, *Social Forces*, 82(4), pp. 1603–1618.

Burawoy, M. (2004c) Public sociologies: A symposium from Boston College: Introduction, *Social Problems*, 51(1), pp. 103–130.

Burawoy, M. (2005) Presidential address: For public sociology, *American Sociological Review*, 70, pp. 4–28.

Carneiro, N. S. (2009) *Homossexualidades: Uma Psicologia Entre Ser, Pertencer e Participar* (Porto: Livpsic).

Carneiro, N. S. & Menezes, I. (2007) From an oppressed citizenship to affirmative identities: Lesbian and gay political participation in Portugal, *Journal of Homosexuality*, 53(3), pp. 65–82.

Cascais, A. F. (Ed.) (2004) *Indisciplinar a Teoria. Estudos gays, lésbicos e queer* (Lisbon: Fenda).

Cascais, A. F. (2006) Diferentes como só nós. O associativismo GLBT português em três andamentos, *Revista Crítica de Ciências Sociais*, 76, pp. 109–126.

Cole, E. R. (2008) Coalitions as a model for intersectionality: From practice to theory, *Sex Roles*, 59(3), pp. 443–453.

Davis, K. (2008) Intersectionality as buzzword: A sociology of science perspective on what makes a feminist theory successful, *Feminist Theory*, 9(1), pp. 67–85.

Della Porta, D. & Diani, M. (1999) *Social Movements. An Introduction* (Oxford: Blackwell).

de Sousa Santos, B. (2006) The university in the 21st century: Towards a democratic and emancipatory university reform, in: R. Rhoads & C. A. Torres (Eds) *The University, State, and Market. The Political Economy of Globalization in the Americas*, pp. 60–100 (Stanford, CA: Stanford University Press).

Durkheim, E. ([1912] 2007) The elementary forms of the religious life, in: C. Calhoun, J. Gerteis, J. Moody, S. Pfaff & I. Virk (Eds) *Classical Sociological Theory*, pp. 181–192 (Oxford: Blackwell).

Engel, S. M. (2001) *The Unfinished Revolution. Social Movement Theory and the Gay and Lesbian Movement* (Cambridge: Cambridge University Press).

Filax, G. (2006) Politicising action research through queer theory, *Educational Action Research*, 14(1), pp. 139–145.

Flacks, D. (2005) A questão da relevância no estudo dos movimentos sociais, *Revista Crítica de Ciências Sociais*, 72, pp. 45–66.

Foweraker, J. (1995) *Theorizing Social Movements* (London: Pluto).

Fraser, N. & Honneth, A. (2003) *Redistribution or Recognition? A Political-Philosophical Exchange* (London: Verso Books).

Gamson, W. (2004) Life on the interface, *Social Problems*, 51(1), pp. 103–130.

Gans, H. J. (2002) More of us should become public sociologists, Available at http://www.asanet.org/footnotes/julyaugust02/fn10.html (accessed 17 December 2009).

Giffney, N. (2004) Denormatizing queer theory: More than (simply) gay and lesbian studies, *Feminist Theory*, 5(1), pp. 73–78.

Giffney, N. & O'Rourke, M. (Eds) (2009) *The Ashgate Research Companion to Queer Theory* (Surrey: Ashgate).

Goodwin, J. & Jasper, J. M. (Eds) (2009) *The Social Movements Reader. Cases and Concepts* (Oxford: Wiley-Blackwell).

Gouldner, A. ([1972] 2004) Toward a reflexive sociology, in: C. Seale (Ed.) *Social Research Methods*, pp. 381–383 (London: Routledge).

Gramsci, A. (1971) *Selections from the Prison Notebooks* (London: Lawrence and Wishart).

Halberstam, J. (2003) What's that smell? Queer temporalities and subcultural lives, *International Journal of Cultural Studies*, 6(3), pp. 313–333.

Haraway, D. (2004) Situated knowledges: The science question in feminism and the privilege of partial perspective, in: S. Harding (Ed.) *The Feminist Standpoint Theory Reader. Intellectual & Political Controversies*, pp. 81–101 (New York: Routledge).

Harding, S. (1991) *Whose Science? Whose Knowledge? Thinking From Women's Lives* (New York: Cornell University Press).

Harding, S. (Ed.) (2004) *The Feminist Standpoint Theory Reader, Intellectual & Political Controversies* (New York: Routledge).

Harding, S. & Norberg, K. (2005) New feminist approaches to social science methodologies: An introduction, *Signs: Journal of Women in Culture and Society*, 30(4), pp. 2009–2015.

Haug, F. (Ed.) (1987) *Female Sexualisation: A Collective Work of Memory* (London: Verso).

Held, V. (2006) *The Ethics of Care: Personal, Political, Global* (Oxford: Oxford University Press).

Irvine, J. (2003) The sociologist as voyeur: Social theory and sexuality research, 1910–1978, *Qualitative Sociology*, 26(4), pp. 429–456.

Lister, R. (1997) *Citizenship: Feminist Perspectives* (Basingstoke: Macmillan).

Medhurst, A. & Munt, S. R. (Eds) (1997) *Lesbian and Gay Studies. A Critical Introduction* (London: Cassel).

Namaste, K. (1996) The politics of inside/out: Queer theory, poststructuralism, and a sociological approach to sexuality, in: S. Seidman (Ed.) *Queer Theory/Sociology*, pp. 194–212 (Oxford: Blackwell).

Nogueira, C. & Oliveira, J. (Eds) (2010) *Estudo sobre a Discriminação em Função da Orientação Sexual e da Identidade de Género* (Lisbon: CIG).

Oakley, A. (1982) Interviewing women: A contradiction in terms, in: H. Roberts (Ed.) *Doing Feminist Research*, pp. 30–61 (London: Routledge).

Oliveira, J. M., Pinto, P., Pena, C. & Costa, Carlos G. (2009) Feminismos queer: Disjunções, articulações e ressignificações, *Ex-Aequo*, 20, pp. 13–27.

Pfohl, S. (2004) Blessings and curses in the sociology classroom, *Social Problems*, 51(1), pp. 103–130.

Roseneil, S. (2004) Why we should care about friends: An argument for queering the care imaginary in social policy, *Social Policy and Society*, 3, pp. 409–419.

Ryan, C. (2004) Can we be compañeros? *Social Problems*, 51(1), pp. 103–130.

Ryan-Flood, R. & Gill, R. (Eds) (2010) *Secrecy and Silence in the Research Process. Feminist Reflections* (London: Routledge).

Sanchez-Taylor, J. & O'Connell Davidson, J. (2010) Unknowable secrets and golden silence, in: R. Ryan-Flood & R. Gill (Eds) *Secrecy and Silence in the Research Process. Feminist Reflections*, pp. 42–53 (London: Routledge).

Santos, A. C. (2005) *A Lei do Desejo. Direitos Humanos e Minorias Sexuais em Portugal* (Porto: Afrontamento).

Santos, A. C. (2006a) Entre a academia e o activismo: Sociologia, estudos queer e movimento LGBT em Portugal, *Estudos Queer: Identidades, Contextos e Acção Colectiva* [special issue], *Revista Crítica de Ciências Sociais*, 76, pp. 91–108.

Santos, A. C. (Ed.) (2006b) *Estudos Queer: Identidades, Contextos e Acção Colectiva* [special issue], *Revista Crítica de Ciências Sociais*, 76.

Santos, A. C. (2008) Enacting activism. The political, legal and social impacts of LGBT activism in Portugal, PhD thesis, University of Leeds, UK.

Santos, A. C. (2011) Vidas cruzadas: Activismo, ciência e interseccionalidade nos Estudos LGBTQ, *Les Online*, 3(1), pp. 24–28.

Seidman, S. (1996) *Queer Theory/Sociology* (Oxford: Blackwell).

Shields, S. (2008) Gender: An intersectionality perspective, *Sex Roles*, 59(3), pp. 301–311.

Skeggs, B. (2004) Exchange value and affect: Bourdieu and the self, in: B. Skeggs & L. Adkins (Eds) *Feminism After Bourdieu*, pp. 75–97 (Oxford: Wiley-Blackwell).

Stanley, L. (1991) Feminist auto/biography and feminist epistemology, in: J. Aaron & S. Walby (Eds) *Out of the Margins: Women's Studies in the Nineties*, pp. 204–219 (London: Falmer).

Taylor, Y., Hines, S. & Casey, M. E. (Eds) (2010) *Theorizing Intersectionality and Sexuality* (Basingstoke: Palgrave MacMillan).

Taylor, Y. & Addison, M. (2011) Placing Research: 'City Publics' and the 'Public Sociologist', *Sociological Research Online*, 16(4).

Touraine, A. (1981) *The Voice and the Eye* (Cambridge: Cambridge University Press).

Valentine, G. (2007) Theorizing and researching intersectionality: A challenge for feminist geography, *Professional Geographer*, 59(1), pp. 10–21.

Widerberg, K. (2008) For the sake of knowledge: Exploring memory-work in research and teaching, in: A. E. Hyle, M. S. Ewing, D. Montgomery & J. S. Kaufman (Eds) *Dissecting the Mundane. International Perspectives on Memory-work*, pp. 113–132 (Lanham, BO: University Press of America).

Wylie, A. (2004) Why standpoint matters, in: S. Harding (Ed.) *The Feminist Standpoint Theory Reader: Intellectual and Political Controversies*, pp. 339–351 (New York: Routledge).

Asking Tough Questions: The Ethics of Studying Activism in Democratically Restricted Environments

SANDRA SMELTZER

Faculty of Information and Media Studies, Western University, London, Canada

ABSTRACT *In this article I examine the ethics of researching media-related activism in democratically restricted environments. Given that research participants in these locales often place a lot on the line to engage in their chosen political pursuits, including enduring government harassment and persecution, the ethics of conducting research about their activism deserves serious critical attention and analysis. Drawing on fieldwork examples from Southeast Asia, particularly Malaysia, I also critically address the safety and welfare of researchers working in restricted milieus and of students interning with politically oriented non-governmental and community-based organizations. Through this discussion I explore what kinds of activism can and should be included under the broad umbrella of activist research, and how academics balance scholarly production expectations with activist commitments on the ground. I contend that activist-oriented research can include a range of complementary hands-on activities, from front-line, direct-action social justice pursuits to less visible, though no less important, "back office" support for local organizations and social movements.*

As a critical communications scholar with a background in anthropology and international development, I am particularly interested in how citizens use information and communication technologies (ICTs) to promote and facilitate democratic ideals and actions. My research focuses predominantly on how politically oriented citizens living in democratically restricted environments employ ICTs to connect with each other, circumvent government media control measures, raise awareness of political issues, pressure local authorities and reach out to transnational supporters. In this article, I explore the ethical terrain of researching such media-related activist endeavours, drawing on examples from long-standing fieldwork experience in Malaysia and other parts of Southeast Asia. I contend that activist-oriented research can include a range of complementary hands-on activities, from direct-action social justice pursuits to less visible "back office" support for local organizations and social movements. While front-line activism is incredibly powerful under optimal conditions, in other contexts it can jeopardize the safety and welfare of both researchers and research participants.

Concomitantly, the back office labour of academics—such as conducting research, offering grant-writing or legal assistance—deserves greater recognition for its contribution to local struggles.

My own activist-oriented research in Southeast Asia is informed by three fields of academic enquiry with a shared goal of integrating theory and concrete political practice. The first is political economy of communication, which is deeply invested in how, why, by whom and to what ends the social activity of communication is owned and controlled by political and economic forces.[1] Scholars in this field are inherently praxis oriented; they call attention to unfair power distributions in the world and are committed to promoting greater equality for all citizens (see Mosco, 2008, 2009). Second, my work is heavily influenced by an academic training in anthropology, with a particular commitment to an 'anthropology in use' that applies 'theories, concepts, and methods from anthropology to confront human problems that often contribute to profound social suffering' (Rylko-Bauer et al., 2006, p. 179). Third, I frame much of my research within development studies, notably the interdisciplinary sub-field of ICTs for development (ICT4D). Critical scholarship in this area explores the role ICTs can play as tools to support and advance socio-political and economic development, especially in marginalized communities around the world (see Servaes, 2008; Kleine & Unwin, 2009).

For many students and some faculty members, this type of international activist-oriented research holds a certain mystique and "sexy" allure. In practice, however, it is never easy, straightforward or stress free. Conducting cross-cultural fieldwork usually requires academics to secure funding, obtain ethics clearance, arrange time away from home and family, and coordinate and conduct interviews with a diverse range of citizens in another locale. In addition to the customary fieldwork requirements—including knowledge of myriad academic literature relevant to one's research area, of regional and country-specific literature, and of various methodological approaches—I earn significant credibility with interviewees if I demonstrate up-to-date and intimate familiarity of what is happening on the ground locally and nationally. Though I am very fortunate to have the freedom to follow my research interests, the behind-the-scenes work required to conduct such research productively must be acknowledged.

Additionally, researchers must navigate issues related to cultural differences in gender expectations, views on sexuality, perceptions of outsiders, human and animal rights, environmental protection, perspectives on democracy and so forth (see Sherif, 2001; Routledge, 2004; Liamputtong, 2010, pp. 123–132). As Paul Routledge writes, during fieldwork researchers 'always negotiate and interact with difference.... this requires a relational ethics of research to be adopted that is sensitive to various degrees and kinds of difference (e.g. gender, ethnicity, age, class, sexuality, etc), but also to the problematic and unequal relations of power that exist between research collaborators' (2004, p. 86). Indeed, I am often acutely aware of my "otherness" (as a Caucasian, North American, female academic) while in Malaysia and other parts of Southeast Asia. Though never will I be an "insider", much of my life is wrapped up in the people, the culture and the politics of another place. The instability of my subjectivity requires continual critical reflexivity during the planning stages of the research, fieldwork and subsequent knowledge production, especially as I move fairly comfortably and often unconsciously between my different roles as outsider, insider, interviewer, friend, political supporter and colleague. Though such '[r]eflexivity does not dissolve ethical tensions', it does, as Cloke et al.

describe, open up 'possibilities for new ethical and moral maps with which to explore ethical terrains more appropriately and more honestly' (2000, p. 133).

While the ethical considerations described above are likely quite familiar to readers, the extensive body of ethics-related literature concerning cross-cultural fieldwork has focused overwhelmingly on research related to acutely marginalized citizens. In these cases, issues of unequal power relations and colonialist research models should rightfully take centre stage (see Liamputtong, 2010, p. 122). This does not, however, diminish the seriousness and complexity of the ethical issues associated with conducting cross-cultural research related to politically oriented citizens who fall into other demographics of society. Many of the individuals I interview in Malaysia, for example, are urban-based, educated, enjoy some level of financial security and tend to be quite comfortable interacting with foreign academics interested in their work, all of which helps diminish, and in some cases negates or reverses, the power hierarchies that understandably concern ethics review boards.[2] Nevertheless, these citizens live, work and volunteer in a country controlled by a repressive government that actively tries to suppress any sort of resistance to the *status quo*. Given that many of these citizens place a lot on the line to engage in their chosen political pursuits, including enduring government harassment and persecution, the ethics of conducting research about their activism deserves serious critical attention and analysis.

This article therefore highlights ethical issues related to five areas that have received relatively little scholarly attention. These include (1) conducting research within democratically restricted environments where research participants are politically oppressed but would not be categorized as acutely marginalized; (2) conducting research about politically oriented ICT usage by citizens seeking to advance democratic principles in such locales; (3) critically addressing the safety and welfare of not only research participants but also of researchers who conduct fieldwork in restricted milieus; (4) taking into consideration the safety and welfare of students interning with politically oriented non-governmental and community-based organizations (NGOs and CBOs); and (5) questioning what kinds of activism should be included under the broad umbrella of "activist research", and how academics can balance scholarly production expectations with on the ground activist commitments.

To address these issues, the following discussion draws on examples from recent fieldwork in Malaysia for which I interviewed a broad range of activists in the country, including members of NGOs and CBOs, political media practitioners and artists, members of various socio-political movements and coalitions, critical academics and opposition politicians.[3] This primary research was comprised largely of 76 semi-structured, open-ended interviews conducted over the course of four months at various junctures in 2009 and 2010, complemented by occasional structured interviews (e.g. with civil servants and government officials) and innumerable casual conversations (e.g. over coffee with individuals who have become close friends and/or colleagues). Many of these interviews were coordinated prior to fieldwork; others were arranged while in country through snowball sampling (see Berg, 2004, pp. 106–162; Liamputtong, 2010, pp. 69–71).

The Debt of Research versus Getting "Good Data"

As is the case in many countries, there is a wide spectrum of how Malaysians choose to carry out their respective political endeavours. Some citizens elect to work largely within the domestic political system to promote change as prudently critical members of government committees or as lobbyists targeting key civil servants about a specific issue. Others promote democratic change without overtly provoking the local authorities. Their activities might include initiating a letter-writing campaign or organizing small information sessions for fellow citizens. Still others see their role as directly challenging the *status quo* through overt political activities, such as coordinating public demonstrations, writing for a critical online newspaper, actively participating in a human rights movement or publicly campaigning for an opposition party candidate. Though internal tensions exist in the debate over which methods and what tools are most effective for promoting change, this multi-pronged approach to promoting democracy and resistance to systems of control is, Weiss (2003) has argued, quite effective within Malaysia's political environment. For the research I conduct, I must consider the ethical implications of asking these busy citizens to give of their time and energy to discuss and reflect upon their political objectives and pursuits (see, e.g. Clifford & Marcus, 1986; Bourgois, 1990; Lofland & Lofland, 1995; Liamputtong, 2008, 2010; Ember & Ember, 2009), especially in an environment where the authorities restrict basic human rights including the freedom of expression, a key component of socio-political and economic development (see Sen, 1999).

Unfortunately, though not surprisingly, all 10 members of the Association of Southeast Asian Nations (ASEAN) consistently rank near the bottom of the Reporters Without Borders Press Freedom Index (RSF, 2010).[4] These rankings indicate that expressive freedoms are seriously lacking in this region of the world as political elites, often in conjunction with other economically powerful actors, try to control how, with whom and about what citizens communicate. The type and extent of these endeavours range from the hegemonic to the more coercive, from requiring media organizations to obtain annual publishing licenses as a means of encouraging self-censorship to arresting political authors, bloggers and online media practitioners.

In Malaysia, I have witnessed the government's extensive efforts to stymie resistance to its long-standing regime. Local authorities publicly harass politically contentious individuals and groups (in large part as a means to dissuade others from following suit), confiscate and ban printed materials deemed politically problematic,[5] and have arrested activists considered especially effective at reaching and influencing fellow citizens—particularly the Malay population, the ruling party's most valuable voting demographic (see, e.g. Anuar, 2008; Sani, 2008). The government has also shut down opposition party papers during a federal election (Smeltzer & Lepawsky, 2010), and has actively—and oft-times forcefully—disrupted and prevented public assembly of a political nature. In July 2011, for example, the police fired water cannons and tear gas at a massive crowd of citizens participating in a peaceful rally demanding electoral reform in the country. Organized by the Coalition for Clean and Fair Elections, or Bersih (meaning 'clean' in Malay), an umbrella organization of 62 NGOs, the 'walk for democracy' drew approximately 20,000–25,000 people to the streets of Kuala Lumpur, roughly a thousand of whom were detained by the authorities (Harvey, 2011; The Guardian, 2011).

Additionally, the government employs a range of laws—notably the draconian Internal Security Act and the Official Secrets Act—to encourage self-censorship, while the Printing Presses and Publications Act (PPPA) stipulates that all presses require a license to print any type of material. The license must be renewed annually and can be revoked or suspended without warning if the government decides that a publication represents a threat to national security (see Laws of Malaysia, 2006). The government announced that it hopes to extend the PPPA to the online world, essentially licensing the Internet, which presumably would include Web 2.0 applications such as Facebook and other social networking sites (see Randhawa, 2011 for an insightful commentary on the implications of such a move).[6]

Given that many other governments around the world employ similar mechanisms to control their own citizens' communicative activities, it is critically important to contextualize the relationship between ICTs and activism (Kling, 2000). Relationships differ and shift depending on the given socio-political milieu; government programmes and policies that promote and restrict certain kinds of technology usage; citizens' access, desire and ability to use particular technologies and the opportunities available for citizens to translate online politics into "real life" change. It is also important not to be determinist about the role ICTs play in actualizing "real" political change. It is, for example, highly misleading to describe the 2011 uprising in Egypt as a Twitter revolution—a technologically determinist narrative that significantly downplays the local socio-political context, including the *longue durée* of labour movements at the heart of the unrest. In like fashion, text messaging did not by itself bring down President Estrada's reign in the Philippines, the fax machine was not the principal political factor in the 1989 Tiananmen Square tragedy and the audiocassette did not instigate the 1979 Iranian Revolution.[7] Within mainstream ICT4D academic and policy literature, there is, however, a strong tendency to veer towards this type of technologically determinist perspective, one that views the presence of new technologies as evidence of improved lives and development writ large without taking into consideration the broader local context. Additionally, we must acknowledge the increasingly invasive surveillance functions of ICTs, and how they are employed by political and economic interests to further commodify our everyday lives, which serve to detract from the technology's democratic potential (see Mosco, 2008; Sandoval & Fuchs, 2010; Fuchs, 2011).

Notwithstanding these limitations, citizens actively employ tools of technology to crack open systems of domination, promote democratic ideals and action, 'raise consciousness, generate public dialogue and debate... and mobilize social action on many fronts' (della Porta *et al.*, 2006; Juris, 2008; Leistyna & Mollen, 2008, p. 26; Atkinson, 2010; Sandoval & Fuchs, 2010). In fact, citizens excluded from traditional channels of communication are often the ones on the cutting edge of technological adaptation and creative usage. In Malaysia, for example, the domestic blogosphere and other Web 2.0 applications—especially Facebook, which became the number one website in the country in 2009 (Alexa, 2010)—have become increasingly popular with citizens wanting to discuss, debate and share political information that hitherto has been unavailable at any mass level because of government controls (Kenyon, 2010; Smeltzer & Keddy, 2010; Smeltzer & Paré, 2010).

Although scholars are beginning to pay more serious attention to this area of research, especially following the early 2011 political uprisings in parts of Northern Africa and the Middle East, to date there has been a relative dearth of critical research addressing the relationship between new forms of ICTs and activist endeavours in repressive states (Atkinson, 2010, p. 153).[8] Yet, political leaders in these countries (including Malaysia) have found themselves in unfamiliar and uncomfortable territory—although they want their citizens to be technologically savvy for competitive economic reasons, they do not want ICTs to be used for political purposes that could challenge their authority and the *status quo*. Since ICTs are used to support, promote and amplify activist activities and that governments are keen to control their political influence, often through whatever means necessary, we must consider seriously the ethical issues associated with conducting this type of research. I am particularly concerned that my interactions with some Malaysians may raise (additional) red flags with the authorities about their political endeavours. Many interviewees are well aware of the risks associated with their pursuits and have, to varying degrees, made the decision to accept such liabilities. Talking to me at a coffee shop is not nearly as contentious as the majority of their other daily activities. The situation would likely change, however, if I interviewed an up-and-coming political blogger or artist who may not be ready for, or completely understand the implications of, increased attention from the authorities.

One particularly salient example from my fieldwork along these lines stands out for me. Several years ago, I interviewed a young man in his mid-20s who was employed at a high-tech, public–private organization in Malaysia. He had also just started to blog, publishing his relatively tame political content under a pseudonym. Our interview focused primarily on issues pertaining to the country's technological future, the effectiveness of government strategies to support the local high-tech sector and promote an IT-savvy citizenry and the potential political role of new media to circumvent state controls. Although this individual signed a consent form, received full disclosure about the research and was guaranteed anonymity, he emailed me a few days after our meeting asking to have his interview removed from my research. I, of course, agreed to his request, but did so without knowing why he had changed his mind. What I find particularly noteworthy about this example is my initial reaction to his email: 'Oh no. I've lost valuable data!' Later, I realized that in reacting this way I had prioritized gathering "good data", and thus my academic production exigencies, over the concerns and safety of an interviewee.

While the research I conduct about political struggles in parts of Southeast Asia is not as traumatic as that undertaken by some of my colleagues (e.g. research that includes personal stories about egregious human rights violations), my interviews often deal with issues and events that negatively impact the lives of other people (e.g. punitive government repression of citizens' freedom of expression). And, yet, I have admittedly felt a sense of giddiness about obtaining such rich material. I am not alone in this feeling—Dickson-Swift *et al.* conducted semi-structured interviews with a range of qualitative public health researchers to gain insight into some of the key challenges involved in research about sensitive subject matter. A number of respondents acknowledged that even though their research is geared primarily towards the betterment of other's lives, the 'sense of excitement' they experience gathering data 'is often in stark contrast with their ethics about "using" people for research purposes' (Dickson-Swift *et al.*, 2007, p. 343). In a similar vein, from 2004 to 2005, Clark conducted an email survey of 55 political scientists carrying out fieldwork in numerous locales throughout the Middle East. All of the participants in the study reported that they felt the same type of 'uneasy debt of interviewees devoting time to the respondents' research' (Clark, 2006, p. 420).

As another relevant example from my Malaysian-based research, a few years ago I spent several hours one day observing and talking to employees and volunteers at a medium-sized, transnational NGO headquartered in Kuala Lumpur. My objective was to learn about the organization and gain a more nuanced understanding of local and international laws relevant to its area of expertise. I quickly realized that all of the individuals at this NGO were highly educated, financially solvent and well connected locally and regionally.

Consequently, while this was an incredibly informative and enjoyable experience, I struggled with how to repay them for their hospitality, time and knowledge sharing. Initially, I had thought to offer them the support of a student intern from my academic institution to assist with their campaign (As discussed in greater detail below, I coordinate undergraduate and graduate student internship placements with various NGOs and CBOs at home and abroad to give young people interested in social justice issues hands-on experience, while also providing organizations with additional labour to help fulfil their important mandates). This specific NGO, however, did not feel that it was equipped to host an internship placement, which meant that I also could not "help" in this manner. While I thanked these individuals for the generosity, I left the NGO feeling like I had been given a valuable gift but could not reciprocate.

Activism and Internships: Throwing Students into the Mix

Over the past several years, I have organized, supervised and secured funding (primarily from my academic institution) for numerous student internships with media-related activist organizations in South and Southeast Asia. Though the organizations I select for internship placements are politically oriented, they are not so contentious as to draw unwanted attention from the authorities. As a case in point, I have not placed interns with a specific NGO that I respect tremendously and where close friends and colleagues volunteer their time and contribute their political capital. Although this organization would likely benefit from the labour of a hard-working, critical media studies student, I am reluctant to coordinate such an internship owing to the more overtly political nature of its work which makes it a target of the government's anti-resistance campaigns. Once again, we see the importance of ethical reflexivity as *I* decide what organizations and what types of politically oriented activities *I* think are appropriate for student involvement. Though I want students to be immersed in the local culture to understand fully the challenges faced by NGOs, CBOs and social movements in the country, I still make judgement calls on their behalf.

I am also very clear that students should not, under any circumstances, join a protest, demonstration or engage in any activity that would attract undue government interest. The students at our institution must sign risk waivers, purchase additional health insurance, attend socio-cultural and political orientation sessions about the country and region in which they will live and work and demonstrate that they are prepared (as much as they can be) emotionally, financially and scholastically to participate in these internships. Despite this preparatory process, and knowing that these students are adults who will hopefully benefit personally and professionally from the experience, I feel responsible for their safety and emotional state while abroad. My concerns are heightened by the fact that students often face 'disturbing socio-economic inequities, language barriers, cultural disparities, and the implications of severely diminished press freedoms and civil liberties' (Wessel, 2007; Smeltzer & Grzyb, 2009, p. 10).

It has been my experience that these internships benefit both parties; however, it is vitally important to recognize that the individuals working and volunteering at host organizations must dedicate significant time and energy to supporting and training our students. While my academic institution is keen to highlight publicly the ways in which it contributes to the local and international "community" (e.g. funding international internships), it does not always recognize that organizations are not simply on the receiving end of "free" labour, nor does it always acknowledge what the organizations give back to and teach our students.[9]

Understanding "Activist" Research

Discussions about cross-cultural research and international internships raise familiar questions about the responsibilities academics have beyond the traditional goals of intellectual production. What role(s) can and should academics play in the lives and societies of the people who are the heart of their research, including in places that are democratically restricted but where research participants are far from powerless? Two common approaches to addressing these questions are the Reciprocity and Activist models (see Carapico, 2006). Examples of token reciprocity actions might include buying coffee, lunch or dinner where culturally appropriate for interviewees, or facilitating introductions between activists with similar democratic goals who could share mutually beneficial resources. At a more passive level, casual conversations and interviews afford a platform for many individuals to talk about what they do and why they do it, which in some cases further validates their work and helps establish solidarity with an extended community.

By comparison, Sheila Carapico describes an activist model as one that advocates 'change within the society being researched... the activist ethos is conscientiously engaged' (Carapico, 2006, p. 430). The forms this activism take can be plotted along a wide spectrum of less to more contentious interventions depending on, among other things, the local socio-political context, concerns about the researcher's safety, his/her relationship with the authorities and local communities and what he/she thinks is the best course of action to effect change. A researcher may help bring political issues to light for the general public by writing for a domestic critical media outlet, participating in and/or helping to organize a public demonstration or a politically infused artistic event, or publicly championing sensitive issues such as land claim struggles and other human rights concerns. While I have engaged in some of these activist endeavours, I have tried to stay clear of many overtly activist, front-line pursuits: ones that might raise red flags for the authorities and thus negatively impact myself and others. I have instead provided organizations and movements with what I would call "back office" support, such as helping with grant writing and proof-reading, conducting background research, coordinating internship placements and providing assistance with new media campaigns.

As a result, I find myself aligned with, but not fully immersed in, Hale's (2006) understanding of activist research. For Hale, activist research is similar to Carapico's model described above insomuch as researchers are committed to advocating change within a particular community. He describes it as 'a method through which we affirm a political alignment with an organized group of people in struggle and allow dialogue with them to shape each phase of the process, from conception of the research topic to data collection to verification and dissemination of the results' (Hale, 2006, p. 97). To this end, Hale positions activist researchers as having a dual responsibility to their academic community and to a specific political struggle. Hale differentiates this research from what he calls 'cultural critique'—an 'approach to research and writing in which political alignment is manifested through the content of the knowledge produced, not through the relationship established with an organized group of people in struggle' (2006, p. 98). While cultural critique is concerned with the plight of marginalized citizens and is deeply concerned about unfair power relations, for Hale it fails to effect real-life change in the lives of others. For this reason, although Hale views both approaches as valuable and complementary, he constructs activist research as inherently more beneficial based on its twofold commitment to intellectual and hands-on political objectives.

The distinction Hale makes between these approaches is based primarily on methodological grounds. He supports social science-based activist research that has at its disposal positivist methods (e.g. geo-mapping for land claims) useful for direct interventions on behalf of subjugated communities struggling against oppressive sources of power. By comparison, Hale sees cultural critique as 'theoretically important but methodologically limited' (2006, p. 107), positing that marginalized communities 'seeking new rights from the state will have little tolerance for cultural critiques produced by well-meaning ethnographers' (Dyrness, 2008, p. 26).

Despite Hale's commitment to collaboration in his activist research approach, Dyrness (2008) and Juris (2008) call attention to the ways in which he separates the researcher (an outsider who can help) from the object of the research (citizens who need the help), a distinction they consider problematic in their call for more collective forms of research that try to dissolve or overcome such demarcations. In like manner, I *try* not to position myself as an outside observer in my own research and have instead established relationships with individuals and groups engaged in political activities, offering various forms of support for their causes. My research is not, however, dedicated to a particular group or community, nor do I collaborate with them in shaping the research process. I have instead focused scholarly attention on specific and more broad-based activist causes, including campaigns that promote press freedom, access to information and public assembly rights, and to multifaceted forms of resistance against free trade agreements (FTAs) and environmental degradation. As well, I would not classify most of my political activities as front-line or direct action; rather, I engage predominantly in back office activism as a means of supporting local NGOs, CBOs and social movements, while also trying to produce scholarly material highlighting social justice issues related to unfair domestic power relations.

Although the front-line activism many academics undertake in support of local struggles is incredibly valuable and effective, it raises two important issues that deserve greater recognition. First, many academics also provide less publicly visible, though no less important, forms of support to organizations and movements. I am particularly interested in highlighting the benefits of the back office labour described above—hands-on support which can, but clearly does not need to be, positivist in nature. Although front-line activism and back office activism are not mutually exclusive, the labour involved in engaging in the latter tends to be under-acknowledged in the literature (perhaps because it does not *appear* to signal the same commitment to a cause). Also, I have found that through engaging in back office work for different organizations throughout Southeast Asia, I have gained invaluable insight into how activism operates on the ground in a comparative setting within and across countries. In so doing, I witness the struggles and successes of a multi-pronged approach to resistance and change, including efforts to champion freedom of expression and other communicative rights.

Second, front-line activism is not always appropriate or safe for both researchers and research participants. In a similar vein to Hale's activist research, Juris contends that to engage in what he calls 'militant ethnography', 'one has to build long-term relationships of commitment and trust, become entangled with complex relations of power, and live the emotions associated with direct-action organizing and transnational networking'

(2008, p. 20). Although I am fully in favour of the first two criteria, direct action is not always possible in many contexts. While Juris's extensive (and impressive) experience with anti-corporate globalization protests in Europe is certainly not without danger, similar encounters in other places around the world—including in various parts of Southeast Asia—can prove especially hazardous. It is to these issues of safety and welfare that I now turn.

The Welfare and Responsibility of Researchers in Democratically Restricted Environments

Much has been written about the safety and ethical concerns associated with conducting fieldwork in locales where the authorities possess extensive surveillance capabilities (traditional and/or high-tech) and employ both hegemonic and coercive forms of control to stymie resistance to the *status quo*. Under these circumstances, asking what may seem like an innocuous question can place both interviewer and interviewee in a precarious situation, thus requiring political astuteness to ensure the safety of both parties. Romano (2006), Sanford & Angel-Ajani (2006), Shahidian (2001) and Wood (2006) offer instructive case studies addressing the ethical and security concerns related to conducting research in conflict zones, including places that may not at first appear particularly volatile. The backlash against local and foreign journalists during the February 2011 uprisings in Egypt is a particularly salient example of how a political situation can shift quickly and in ways unexpected (CBC, 2011). Likewise, Jacobsen and Landau (2003), Liamputtong (2010, pp. 51–55) and Mackenzie *et al.* (2007) examine some of the key concerns associated with conducting especially sensitive cross-cultural research and fieldwork in crisis situations (e.g. during a humanitarian emergency).

Although my research in Malaysia is not as dangerous as conflict zone or crisis-situation fieldwork, the government's control measures can impact my ability to explore, critically examine and support activist pursuits. For instance, my interactions with certain local activists are well known to the authorities. I have conducted interviews and engaged in informal conversations with friends and colleagues in public locations, knowing fully well that a member of the Special Branch, Malaysia's intelligence agency, is purposively within earshot. On one occasion, a political blogger relayed to me an experience of being questioned by police in relation to content published on his/her blog. This individual was 'requested' to look at albums of photographs tracking his/her daily activities. Included were pictures of the two of us talking over coffee in three different public locations in Kuala Lumpur. While this kind of experience is rather benign, it does carry with it a somewhat ominous threat of what could happen if I chose to cross-specific political lines. If, let us say, I were to openly and *publicly* question or dispute Malaysia's Bumiputra affirmative action policies, which benefit the majority Malay population to the detriment of other citizens in the country—notably the large Chinese, Indian and Indigenous populations—I would likely move further onto the government's radar. Though it is doubtful that the government would formally or informally reprimand me for writing something meant primarily for the "usual suspects" (e.g. other academics, already-critical NGOs and media practitioners), if I wrote a critical op-ed piece about these policies for a domestic newspaper that could reach the masses, things would almost certainly change for the worse.

Engaging in activist research in other Southeast Asian countries can unfortunately prove more perilous, as Alan Shadrake, a Malaysian-based British writer, former journalist, and author, has experienced. In his book, *Once a Jolly Hangman: Singapore Justice on the Dock*, Shadrake questions the impartiality of Singapore's judiciary vis-à-vis its application of the country's death penalty. After the 2010 launch of his book in Singapore, the 76-year-old Shadrake was arrested, fined and sentenced to jail time (BBC, 2010). While I do not fear a backlash similar to the one experienced by Shadrake, I have had glimpses of what might transpire if I shifted my research priorities. For instance, as a minor part of my larger research agenda from 2007 to 2009, I explored the impact of FTAs on local cultural and physical environments in Southeast Asia. During interviews with various Malaysian-based community organizations and coalitions fighting against FTAs, a few individuals gently but firmly suggested that if I chose to extend my research, I ought to be careful who I talk to about environmental issues in certain regions of the country where illegal logging is expedited by especially powerful actors. They were very clear that I would be jeopardizing my personal safety if I asked questions that even hinted at environmental impropriety. Heeding their advice, I decided not to continue my research in these areas. I wonder, though, if in my younger years, perhaps as a doctoral student hoping to "prove" my fieldwork acuity and/or not fully understanding the potential consequences of my actions, I would have tried to push such boundaries.

This example highlights the marked importance of reflexivity while conducting research in more politically fraught locales, as decisions must be made that balance safety concerns with obtaining "good data". In the literature concerning qualitative research methodology, discussions about ethical issues are overwhelmingly—and rightfully—focused on protecting research participants. As Liamputtong writes, '[e]thics is a set of moral principles that aim to prevent research participants from being harmed by the researcher and the research process' (2010, p. 32). Nevertheless, the safety and welfare needs of the researchers themselves are 'often thought through in a cursory manner or in an *ad hoc* contingent fashion once in the field' (Lee-Treweek & Linkogle, 2000, p. 1). Hence, the sagacity of David Romano's advice: prior to fieldwork, '[i]t is important to have clear "red-lines" in mind, because once you are in the field there is a natural reluctance to cancel and leave before you originally intended, which can lead to a psychological minimizing of very real and increasing threats' (2006, p. 440).

Given that much of my work critiques the very institutions with the power to quash resistance to their authority, engaging in politically contentious activist research and activities—especially those that are publicly visible—could potentially place me and others in a dangerous situation and hinder my ability to conduct future work in the area. At the same time, due to my privileged place as a Western-based academic, how can I not engage in Hale's form of activist research? How can I not try, through direct action, to make a concrete difference in the lives of the people who are at the heart of my research? As Routledge writes, academics are 'entangled within broader powers of association and intellectual production' that grant them 'certain securities and advantages – for example economic, political, representational – that may not be enjoyed by those with whom we collaborate – especially if they live and work beyond the academy' (2004, p. 84; see also Fuller & Kitchin, 2004; Liamputtong, 2010, p. 225). Likewise, I believe that academics have an important responsibility to act as public and engaged intellectuals beyond the "ivory tower" (see Fuller & Kitchin, 2004; Coté *et al.*, 2007; Smeltzer & Grzyb, 2009). As critical pedagogy scholar Henry Giroux has argued passionately, '[a]cademics can no longer retreat into their careers, classrooms, or symposiums as if they were the only spheres available for engaging the power of ideas and the relations of power' (1991, p. 57).

However, in addition to safety considerations, it is important to recognize—and be willing to admit—that the political interventions researchers make in other locales may not always be appropriate or helpful even when they are collaborative (see Cahill & Torre, 2007). While my work is categorically not value-free (i.e. it does not pivot on cultural relativism and is principally concerned with promoting democracy writ large), I do, as noted, make choices about who to interview, what to ask and so forth. These choices are tied directly to how I view social justice, what I consider to be the most effective means of improving democracy and what my role can and should be in the process (see Martin, 2007; Butz, 2008). For these reasons, I must have a strong grasp of the local socio-political, cultural and economic context of my research locale to understand the potential ramifications of my actions. Indeed, we as academics 'need to acknowledge that we cannot see into the future to know what are the long-term implications of our research practices on research participants' lives as well as our own' (Routledge, 2004, p. 87).

As well, the hefty pedagogical, research and university service demands of our jobs, combined with personal and family commitments, leave little time, room and energy for the kinds of activist work many of us *want* to undertake. In particular, the publish-or-perish expectations that bear down on academics in the quest for promotion and tenure 'mean that outreach, engagement, and community service usually do not figure strongly in the adjudication of a faculty member's productivity' (Ward, 2005, p. 219; see also Few *et al.*, 2007; Hale, 2008).

Criticality versus Censorship

As academics navigate their scholarly and activist commitments, they are faced with what Routledge calls the dilemma of 'criticality versus censorship': 'how critical can one be and still continue to support rather than undermine a particular struggle?' (2004, p. 87). A director of an NGO, a volunteer at a critical media outlet or a coordinator of a human rights movement is unlikely to take kindly to being critiqued, especially by a foreign researcher. As such, without consciously realizing it, I think I have tempered my analysis of some of the activist work undertaken by the organizations that host our interns, wanting to maintain a positive relationship and working environment for the students. I have also felt pangs of apprehension about how the criticality versus censorship conundrum plays out in the 'politics of knowledge production' (Speed, 2006, p. 71), including the implications of 'publishing conclusions that are either unflattering to the interviewees or contradict the interviewees' opinions' (Clark, 2006, p. 420; Sultana, 2007).

The case of the transnational NGO headquartered in Malaysia discussed earlier offers insight into this knowledge production conundrum. Approximately, a year after I conducted a semi-structured interview with an employee of this NGO, I published a scholarly piece that included a gentle critique of the effectiveness of some of the practices undertaken by domestic NGOs and CBOs in their quest to raise awareness about a specific set of political issues. The publication also recognized the valuable contributions these organizations make to supporting local democratic efforts, especially given the challenges of operating within such a repressive environment. Nevertheless, the next time I was in the country, I contacted this individual hoping to arrange another coffee or lunch, an invitation that was politely declined without a reason being offered. While there may be a range of explanations for the refusal, I know members of this organization monitor academic material related to their work and thus may have come across my critique and chosen not to entertain future interviews. Similar to many other academics, I have found it difficult to strike a meaningful balance between supporting the work of activists like this who are dedicated to improving the lives of others, and producing a theoretically informed *and critical* analysis of their perspectives and actions. Although these two objectives are not at odds with one another, they do require a thoughtful and honest examination of what kinds of scholarly material I produce. There are also valid concerns about who else could read such material and to what ends they might use the information therein. As Routledge advises, 'writing about resistance formations in scholarly journals needs to tread a fine line between support for a struggle and the professional and ethical requirements to be constructively critical while also not providing help to the opponents of that struggle' (2004, p. 87; see also Clark, 2006, p. 420; Carapico, 2006).

Re-categorizing Activist Research

Through my research experiences in democratically restricted environments, it has become abundantly clear to me that theory and practice are not mutually exclusive. In fact, I consider the exact opposite to be the case—the theoretical perspicacity I gain via political praxis is invaluable for my scholarly productivity, while my theoretical training has provided a more nuanced understanding of the issues faced by citizens engaged in activist activities. Hale writes, and I concur, that:

> To align oneself with a political struggle while carrying out research on issues related to that struggle is to occupy a space of profoundly generative scholarly understanding. Yet when we position ourselves in such spaces, we are also inevitably drawn into the compromised conditions of the political process. The resulting contradictions make the research more difficult to carry out, but they also generate insight that otherwise would be impossible to achieve. This insight, in turn, provides an often unacknowledged basis for analytical understanding and theoretical innovation (2006, p. 98).

Similarly, Shannon Speed defends the 'productive tensions' produced through 'critically engaged activist research' that brings together theory, critical analysis and 'concrete political objectives' (2006, p. 71). We must, however, acknowledge that the 'compromised conditions of the political process' can be quite serious in certain locales and with regards to politically sensitive issues. Drawing on my three areas of theoretical interest, as explained in the introduction, much of my work explores the opportunities and constraints that influence citizens' capability to leverage tools of technology to actualize democratic activities *they* believe will lead to a better future for themselves and their fellow citizens (Robeyns, 2005; Smeltzer & Paré, 2010). In critically examining this line of enquiry in places where the authorities are not keen to entertain development objectives that differ from their own, ethical reflexivity is paramount for protecting the welfare of both researcher and research participants.

Equally important to recognize is the wide spectrum of activist undertakings available to academics who want to support local struggles, including labour that is neither front-line nor publicly demonstrative. As noted above, the multi-pronged approach to activism in Malaysia has proven quite effective—citizens draw on a range of tools and methods, working both inside and outside the system, to promote democratic change in a society controlled by a semi-authoritarian regime. In like fashion, activist-oriented researchers employ a variety of tools and methods—some or all of which may be out of the public eye—that are beneficial to NGOs, CBOs and social movements, and which help build the 'long-term relationships of commitment and trust' (Juris, 2008, p. 20) necessary for other types of direct-action activism when and where appropriate. To be clear, the goal here is not to be an apologist for back office activism at the expense of more direct-action involvement. I instead want to highlight the array of activist pursuits available to academics, while also recognizing the circumstances that can limit more direct forms of activism.

Indeed, in both their hands-on pursuits and scholarly production, academics must navigate the murky ethical waters of criticality versus censorship, and of balancing 'activism with the ethical responsibilities that accrue to being a representative of an academic institution' (Routledge, 2004, p. 87). These balancing acts are always context specific and must take into consideration the very real—and often overlooked—safety and security concerns associated with conducting certain types of activist research that could attract unwanted attention from local authorities. While much of this article references the work of anthropologists, other disciplines, including critical media and communication studies, would do well to focus greater attention on such important issues of reflexivity vis-à-vis activist research.

Notes

1. This field includes a fairly wide range of perspectives. Dwaye Winseck (2011) offers an especially useful overview of four key schools of thought—Neoclassical Economics, Radical Media Political Economy, Schumpeterian Institutional Political Economy and the Cultural Industries School—that fall under the rubric of political economy.
2. Yet, institutional research ethics boards tend to favour a clear-cut, and some would argue apolitical, research agenda prior to the commencement of one's fieldwork (see Bourgois, 1990; Blake, 2007; Martin, 2007). Though the academic ethics review process is essential for helping to protect the welfare of research participants, it has become so 'codified' and 'standardized' (Blake, 2007, p. 413) that it often fails to accommodate the flexibility required to conduct qualitative fieldwork. This flexibility is essential for researchers who routinely negotiate *in situ* what is and is not appropriate to ask interviewees, what general issues are socially and politically acceptable and safe as topics of discussion, who should be interviewed and where and how participant observation should take place. As well, the process tends to construct research participants as objects 'upon which research is done' (Blake, 2007, p. 414), thereby framing them as powerless, which does not reflect the vast majority of my research experience. Of note, this is not, however, meant to suggest that acutely marginalized research participants are, by comparison, without power.
3. These groups are not though mutually exclusive as politically contentious Malaysians are often affiliated with more than one organization within this community.
4. Of the 178 countries listed in the 2010 index, Myanmar is the lowest ranking Southeast Asian country in the 174th spot. The other nine countries, in descending order, include Indonesia (117), Cambodia (128), Singapore (136), Malaysia (141), Brunei (142), Philippines (156), Thailand (153), Vietnam (165) and Laos (168) (RSF, 2010).
5. A well-publicized and recent example is the government's decision to ban the political cartoon books of Zulkifli Anwar Ulhaque, commonly referred to by his penname Zunar. In the fall of 2010, Zunar was arrested and charged with sedition just prior to the launch of his book, *Cartoon-O-Phobia* (personal communication, 2010).
6. For a regularly updated overview of the constraints placed upon freedoms of press, speech, expression and assembly in Malaysia, see the Center for Independent Journalism (www.cij.org), Human Rights Watch (www.hrw.org), Reporters without Borders (www.rsf.org) and the Southeast Asian Press Alliance (www.seapabkk.org).

7. Be that as it may, I frequently interview members of Malaysia's (broadly speaking) civil society who give significant—and I contend unwarranted—credit to technology itself for positive democratic shifts in their country. In doing so, they downplay much of their important political work and the work of their fellow citizens (Smeltzer & Lepawsky, 2010).

8. The critical content citizens produce and consume via ICTs is referred to by scholars in different fields as activist, alternative, critical, independent, radical and community-oriented, as debates ensue over how to classify such non-mainstream forms of media and whether to do so according to their content, format and/or funding mechanisms (Waltz, 2005; Bailey *et al.*, 2008; Pajnik & Downing, 2009).

9. See van't Klooster *et al.* (2008) for an insightful discussion of overseas internship programmes, especially in developing and newly industrializing countries.

References

Alexa Internet (2010) Top sites in Malaysia. Available at http://www.alexa.com/topsites/countries/MY (accessed 21 November 2010).

Anuar, M. K. (2008) Media commercialisation in Malaysia, in: C. George (Ed.) *Free Markets, Free Media? Reflections on The Political Economy of the Press in Asia*, pp. 124–136 (Singapore: AMIC press).

Atkinson, J. D. (2010) *Alternative Media and Politics of Resistance: A Communication Perspective* (New York: Peter Lang).

Bailey, O., Cammaerts, B. & Carpentier, N. (2008) *Understanding Alternative Media* (Maidenhead, England; New York: McGraw Hill/Open University Press).

BBC (2010) UK Author Shadrake Jailed for Six Weeks in Singapore, 16 November 2010. Available at http://www.bbc.co.uk/news/world-asia-pacific-11763031 (accessed 18 November 2010).

Berg, B. (2004) *Qualitative Research for the Social Sciences*, 5th ed. (Boston, MA: Pearson Education, Inc.).

Blake, M. K. (2007) Formality and friendship: Research ethics review and participatory action research, *ACME: An International E-Journal for Critical Geographies*, 6(3), pp. 411–421.

Bourgois, P. (1990) Confronting anthropological ethics: Ethnographic lessons from central America, *Journal of Peace Research*, 27(1), pp. 43–54.

Butz, D. (2008) Sidelined by the guidelines: Reflections on the limitations of standard informed consent procedures for the conduct of ethical research, *ACME: An International E-Journal for Critical Geographies*, 7(2), pp. 239–259.

Cahill, C. & Torre, M. E. (2007) Beyond the journal article: Representations, audience, and the presentation of participatory research, in: S. Kindon, R. Pain & M. Kesby (Eds) *Connecting People, Participation and Place: Participatory Action Research Approaches and Methods*, pp. 196–205 (London: Routledge).

Carapico, S. (2006) No easy answers: The ethics of field research in the Arab World, *PS: Political Science & Politics*, 39(3), pp. 429–431.

CBC (2011) Foreign Journalists Attacked in Egypt, 3 February 2011. Available at http://www.cbc.ca/news/world/story/2011/02/03/egypt-journalists.html (accessed 6 February 2011).

Clark, J. A. (2006) Field research methods in the Middle East, *PS: Political Science & Politics*, 39(3), pp. 417–424.

Clifford, J. & Marcus, G. (1986) *Writing Culture: The Poetics and Politics of Ethnography* (Berkeley, CA: University of California Press).

Cloke, P., Cooke, P., Cursons, J., Milbourne, P. & Widdowfield, R. (2000) Ethics, Reflexivity and research: Encounters with homeless people, *Ethics, Place & Environment*, 3(2), pp. 133–154.

Coté, M., Day, R. J. F. & de Peuter, G. (Eds) (2007) *Utopian Pedagogy: Radical Experiments Against Neoliberal Globalization* (Toronto, Buffalo, London: University of Toronto Press).

della Porta, D., Andretta, M., Mosca, L. & Reiter, H. (2006) *Globalization From Below: Transnational Activists and Protest Networks* (Minneapolis, MN: University of Minnesota Press).

Dickson-Swift, V., James, E. L., Kippen, S. & Liamputtong, P. (2007) Doing sensitive research: what challenges do qualitative researchers face?, *Qualitative Research*, 7(3), pp. 327–353.

Dyrness, A. (2008) Research for change versus research as change: Lessons from a *Mujerista* participatory research team, *Anthropology & Education Quarterly*, 39, pp. 23–44.

Ember, C. R. & Ember, M. (2009) *Cross-Cultural Research Methods*, 2nd ed. (Plymouth: AltaMira Press).

Few, A. L., Piercy, F. P. & Stremmel, A. J. (2007) Balancing the passion for activism with the demands of tenure: One professional's story from three perspectives, *NWSA Journal*, 19(3), pp. 47–66.

Fuchs, C. (2011) New media, web 2.0 and surveillance, *Sociology Compass*, 5(2), pp. 134–147.

Fuller, D. & Kitchin, R. (Eds) (2004) *Radical Theory, Critical Praxis: Making a Difference Beyond the Academy?* Praxis E-Press. Available at http://www.praxis-epress.org/rtcp/fpages.pdf (accessed 28 March 2011).

Giroux, H. A. (1991) *Postmodernism, Feminism and Cultural Politics* (Albany, NY: State University of New York Press).

Hale, C. (2006) Activist research vs. cultural critique: Indigenous land rights and the contradictions of politically engaged anthropology, *Cultural Anthropology*, 21, pp. 96–120.

Harvey, R. (2011) Breezes of Change in Malaysia, *BBC*, 13 July 2011. Available at http://www.bbc.co.uk/news/world-asia-pacific-14138270 (accesses on 16 July 2011).

Jacobsen, K. & Landau, L. B. (2003) The dual imperative in refugee research: some methodological and ethical considerations in social science research on forced migration, *Disasters*, 27(3), pp. 185–206.

Juris, J. (2008) *Networking Futures* (Durham, NC: Duke University Press).

Kenyon, A. T. (2010) Investigating chilling effects: News media and public speech in Malaysia, Singapore, and Australia, *International Journal of Communication*, 4, pp. 440–467.

Kleine, D. & Unwin, T. (2009) What's new about ICT4D?, *Third World Quarterly*, 30(5), pp. 1045–1067.

Kling, R. (2000) Social informatics: A new perspective on social research about information and communication technologies, *Prometheus*, 18(3), pp. 245–264.

Laws of Malaysia (2006) Printing Presses and Publications Act 1984. Available at http://www.agc.gov.my/Akta/Vol.%207/Act%20301.pdf (accessed 20 February 2011).

Lee-Treweek, G. & Linkogle, S. (2000) Overview, in: G. Lee-Treweek & S. Linkogle (Eds) *Danger in the Field: Risk and Ethics in Social Research*, pp. 1–7 (London: Routledge).

Leistyna, P. & Mollen, D. (2008) Teaching social class through alternative media and by dialoging across disciplines and boundaries, *Radical Teacher*, 81(1), pp. 20–27.

Liamputtong, P. (2008) Doing research in a cross-cultural context: Methodological and ethical challenges, in: P. Liamputtong (Ed.) *Doing Cross-Cultural Research: Ethical and Methodological Perspectives*, pp. 3–20 (Dordrecht: Springer).

Liamputtong, P. (2010) *Performing Qualitative Cross-Cultural Research* (Cambridge: Cambridge University Press).

Lofland, J. & Lofland, L. (1995) *Analyzing Social Settings: A Guide to Qualitative Observation and Analysis* (Belmont, CA: Wadsworth).

Mackenzie, C., McDowell, C. & Pittaway, E. (2007) Beyond 'Do No Harm': The challenge of constructing ethical relationships in refugee research, *Journal of Refugee Studies*, 20(2), pp. 299–319.

Martin, D. G. (2007) Bureaucratising ethics: Institutional review boards and participatory research, *ACME: An International E-Journal for Critical Geographies*, 6(3), pp. 319–328.

Mosco, V. (2008) Current trends in the political economy of communication, *Global Media Journal Canadian Edition*, 1(1). Available at http://www.gmj.uottawa.ca/currentissue_e.html (accessed 10 January 2011).

Mosco, V. (2009) *The Political Economy of Communication*, 2nd revised ed., pp. 45–63 (London: Sage), http://www.gmj.uottawa.ca/0801/inaugural_mosco.pdf.

Pajnik, M. & Downing, J. D. H. (Eds) (2009) *Alternative Media and the Politics of Resistance: Perspectives and Challenges* (Ljubljana: Peace Institute).

Randhawa, S. (2011) The impossibility of policing the web, *The Nut Graph*. Available at http://www.thenutgraph.com/the-impossibility-of-policing-the-web/ (accessed 15 March 2011).

Reporters Without Borders (2010) Press Freedom Index 2010. Available at http://en.rsf.org/press-freedom-index-2010,1034.html (accessed 14 March 2011).

Robeyns, I. (2005) The capability approach: A theoretical survey, *Journal of Human Development*, 6(1), pp. 93–114.

Romano, D. (2006) Conducting research in the Middle East's conflict zones, *PS: Political Science & Politics*, 39(3), pp. 439–441.

Routledge, P. (2004) Relational Ethics of Struggle, Available at http://www.praxis-epress.org/rtcp/pr.pdf (accessed 12 February 2011).

Rylko-Bauer, B., Singer, M. & van Willigen, J. (2006) Reclaiming applied anthropology: Its past, present, and future, *American Anthropologist*, 108(1), pp. 178–190.

Sandoval, M. & Fuchs, C. (2010) Towards a critical theory of alternative media, *Telematics and Informatics*, 27, pp. 141–150.

Sanford, V. & Angel-Ajani, A. (2006) *Engaged Observer: Anthropology, Advocacy, and Activism* (New Brunswick, NJ: Rutgers University Press).

Sani, M. A. M. (2008) Freedom of speech and democracy in Malaysia, *Asian Journal of Political Science*, 16(1), pp. 85–104.

Sen, A. (1999) *Development as Freedom* (New York: Random House).

Servaes, J. (2008) *Communication for Development and Social Change* (New Delhi: Sage).

Shahidian, H. (2001) To be recorded in history: Researching iranian underground political activists in exile, *Qualitative Sociology*, 24(1), pp. 55–81.

Sherif, B. (2001) The ambiguity of boundaries in the fieldwork experience: Establishing rapport and negotiating insider/outsider status, *Qualitative Inquiry*, 7(4), pp. 436–447.

Smeltzer, S. & Grzyb, A. (2009) Critical media pedagogy in the public interest, *Democratic Communiqué*, 23(2), pp. 1–22.

Smeltzer, S. & Keddy, D. (2010) Won't you be my (political) friend? The changing face (book) of socio-political contestation in Malaysia, *Canadian Journal of Development Studies*, 30(3–4), pp. 421–440.

Smeltzer, S. & Lepawsky, J. (2010) Foregrounding technology over politics? Media framings of federal elections in Malaysia, *Area*, 42(1), pp. 86–95.

Smeltzer, S. & Paré, D. (2010) The knowledge labour of ICT4D: Wither the separation of carriage and content?, *Ephemera*, 10(3/4), pp. 390–405.

Speed, S. (2006) At the crossroads of human rights and anthropology: Toward a critically engaged activist research, *American Anthropologist*, 108(1), pp. 66–76.

Sultana, F. (2007) Reflexivity, positionality and participatory ethics: Negotiating fieldwork dilemmas in international research, *ACME: An International E-Journal for Critical Geographies*, 6(3), pp. 374–385.

The Guardian (2011) Malaysia police detain hundreds at rally, 9 July 2011. Available at http://www.guardian.co.uk/world/2011/jul/09/malaysia-opposition-protests-elections (accessed 16 July 2011).

Ulhaque, Z. A. (2010) Discussion in Kuala Lumpur, Malaysia, Personal Communication, 26 July 2010.

van't Klooster, E., van Wijk, J., Go, F. & van Rekom, J. (2008) Educational travel: The overseas internship, *Annals of Tourism Research*, 35(3), pp. 690–711.

Waltz, M. (2005) *Alternative and Activist Media* (Edinburgh: Edinburgh University Press).

Ward, K. (2005) Rethinking faculty roles and rewards for the public good, in: A. J. Kezar, T. C. Chambers & J. Burkhardt (Eds) *Higher Education for the Public Good: Emerging Voices from a National Movement*, pp. 217–233 (San Francisco, CA: Jossey-Bass).

Weiss, M. L. (2003) Civil society and political reform in Malaysia, in: D. C. Schak & W. Hudson (Eds) *Civil Society in Asia*, pp. 59–72 (Aldershot: Ashgate).

Wessel, N. (2007) Integrating service learning into the study abroad program: U.S. sociology students in Mexico, *Journal of Studies in International Education*, 11, pp. 73–89.

Winseck, D. (2011) The political economies of media and the transformation of the global media industries: An introductory essay, in: D. Winseck & D. Yong Jin (Eds) *The political Economies of Media: The Transformation of the Global Media Industries* (London: Bloomsbury Academic).

Wood, E. J. (2006) The ethical challenges of field research in conflict zones, *Qualitative Sociology*, 29(3), pp. 373–386.

RESEARCH NOTE

A Personal Reflection on Negotiating Fear, Compassion and Self-Care in Research

S.J. CREEK

Department of Sociology, Hollins University, Roanoke, VA, USA

ABSTRACT *With this paper, I discuss the unexpected, the corporeal and the emotional. I attend not only to the issues of fear and integrity in research of the ex-gay movement in North America, but also to related issues of compassion, self-care and violence. I explore my commitment to communicate empathically with and think compassionately about a group of people who are politically and ideologically quite different from me. In the process of shifting my approach, I began to doubt my ability to write critically about the topics at hand. I faced significant cognitive dissonance as I shared moments of celebration with people who had found relief in the movement. Finally, as I internalized the thousands of pages of ex-gay literature, the dozens of face-to-face interviews and the countless email and instant message exchanges, I began to seriously ponder the violence I had done to myself in the name of research.*

I am backed into the corner of a room. A man with neatly combed hair, khaki pants and a pressed, blue dress shirt is facing me. He is ex-gay, and he is screaming. 'Why', he yells, spit flying in my face, 'are you so far behind on your research of the ex-gay movement? Why haven't you sent me any of your work for me to review? What are you hiding?' He threatens to withdraw his participation if I do not cough up a chapter of my dissertation. I tremble as I offer him inadequate explanations for my lack of productivity. I wake in a cold, profuse sweat.

The fear underlying this dream has been my constant companion during the last four years of studying the ex-gay movement—a religious movement in the United States whose goal is to help the world find 'freedom from homosexuality'. My fear is linked to a number of dilemmas faced during this research. While attending ex-gay conferences, spending time at ex-gay ministries, reading ex-gay literature and interviewing ex-gay persons, I have had to negotiate my own identities and stories as a queer, atheist, feminist, activist, sociologist with the identities and stories of the movement and its people. I was under-prepared for both the intensity of emotions triggered by this research as well as the *emotional labor* required (Hochschild, 1983). Never in my methodological training had it been suggested that I might need to take care of *myself* throughout the research process. In this essay, I attend to the emotional: particularly fear, compassion and doubt in my own research. I explore the risks of what Hubbard et al. (2001) call 'over-empathizing', and I ask: when can research be a type of violence against the self?

Fear and (Self) Loathing at an Ex-Gay Conference

I was nervous when I arrived at the center that would host the annual international ex-gay conference. I registered as a 'reporter/researcher', but as I walked into the building I appeared, hopefully, to be an average attendee. I wanted to throw up. What would people see when they saw me? Would they read me as the queer atheist I really was?

This conference, hosted by Exodus (the largest organization within the movement), was my first exposure to the ex-gay world, beyond websites and books. I had never connected with flesh-and-blood ex-gay persons before—especially in an explicitly ex-gay space. As each day passed, I became increasingly troubled by my internal reactions to the conference, its presenters and the discourse. I worried about how to balance emotional reactions with the research at hand.

In the sessions, I learned about the 'roots' of homosexuality, how to 'journey out of lesbianism', how to reclaim my 'true femininity', why Southern Baptist congregations need to discuss the dangers of homosexuality at the pulpit and the best arguments to counter 'pro-gay theology'. Most of the conference was a blur. The only constant: fear. Ponticelli's essay, *Shades of Grey or Back to Nature? The Enduring Qualities of the Ex-gay Movement*, describes the terror she experienced during her research at an ex-gay conference in the early nineties:

> After scrubbing myself red in the shower, trying to remove every essence of Exodus, I counted the planes while trying to focus on the motorcycles on ESPN. I arrived at the airport at 6 a.m. sharp, hoping to catch an early flight … I never felt out of 'their' grasp … Exodus had introduced me to a set of emotions so fluid that I often felt schizophrenic. (Ponticelli, 2005, p. 154)

Ponticelli put to words experiences I had not been able to verbalize. My fear, I recognized, was not rational. The paranoia, disgust, sadness and doubt plagued me throughout my time with Exodus. On top of these feelings, I carried shame and disappointment in myself. As a researcher, I took meticulous notes each conference day. As a human being, though, I struggled with my feelings. I simply had not expected such a strong reaction. Coming back to my room each night, I looked cagily over my shoulder before locking myself in.

Meeting Fear with Compassion

It took time to recover. I buried myself in the hoops and hurdles of graduate work. When I reflected on the fear I had felt over the summer, I viewed it as a trial by fire from which I had emerged intact. By leaping in head first and attending the conference, I reasoned it made sense that I had experienced an intense, internal reaction.

The following spring, I began advertising on listservs. I emailed leadership around the country, asking them to pass on my request for interviews to their participants. As I corresponded with leaders, I was told again and again that they had been 'burned' by researchers and journalists, and that they were not interested in exposing the persons in their programs to such experiences. I reflected on my own reactions at the conference.

Would I burn these people if I had a chance to talk to them? Would my work be exploitative and overly critical?

As I began my second round of research—driving around North America, talking to ex-gay participants—I vowed to attend to two things: taking better care of myself, and breaking down the re-surfacing fears triggered by this research. I read Marshall Rosenberg's book *Nonviolent Communication: A Language of Life* (1998) with the intention of integrating empathy into the evaluative process of semi-structured interviewing. I wanted to engage with people whom I had previously understood as oppositional. It was a vulnerable sensation. With each interview, I felt exposed, even though I was rarely asked to share about myself. In retrospect, I think I used empathy to overcompensate for the fear and loathing I felt. This sensation was akin to what Hubbard et al. (2001) deem 'over-empathizing', and my ability to sustain a professional detachment was compromised.

Toward the end of my travels, I visited an ex-gay organization in the southeastern United States. 'Yvonne', the leader, was excited by the opportunity for persons in her ministry to share their experiences. Initially, I was told there would be eight interviews, and so I gave myself four days to visit. When I arrived, Yvonne told me she had scheduled 17 interviews, and she gave me a sheet listing each interview time, place and interviewee pseudonym. I was thankful someone handed me 17 ex-gay interviews but frustrated that every detail was orchestrated by the organization's leader. I was worried about people being 'handpicked' to tout the party line, the confidentiality of the people speaking to me (given leadership involvement) and I feared that conducting the interviews in the organization's building might hamper storytelling. I had no idea how I could sustain eight hours of interviewing a day, four days straight. My vow to practice self-care was compromised, but I was in no position to bargain.

The women and men with whom I spoke were likeable and kind; they told me tales of survival and hope. I listened to stories about suicide attempts, the emptiness of their earlier lives, as well as the joy felt as they reconnected with God and found help for their primary issue (same-sex attractions) and secondary issues (suicide, co-dependency, depression, isolation, addiction). I had let go of much of my loathing and pity. Several times, I was told that 'it feels good to talk about this'. I often felt more like a counselor and less like a sociologist, though—a dangerous position for both qualitative researchers and the subjects involved (Burr, 1996; Sampson et al., 2008).

Sustaining empathy took every ounce of my energy. I returned to my hotel room exhausted, each night. I was coping with 'compassion fatigue' (Pickett et al., 1994) to the best of my abilities. Unlike the previous summer, the fear was manageable. My need to attend to my own identities had fallen away, or perhaps I was too exhausted to care.

The Price of Empathy

At times, I questioned my ability to effectively carry out this study. I oscillated between anger at the movement and an empathic connection with subjects. As I shifted from data collection to data analysis, I felt the consequences of my 'over-empathizing'. It was not my own position as a queer atheist that put my judgment into question; it was the connection I felt to my research participants. Many of the persons with whom I spoke hoped my research would 'reach other people who are struggling'. I was frustrated in my analytical process, frozen by a sense that I could not do justice to subjects. I was further conflicted knowing these stories, though strangely dear to me, were antithetical to my own values. As a sociologist and a graduate student, I needed to write a dissertation—a constructionist consideration of the meaning making that happens in the stories people tell about their identities. I was unable to give my participants a voice that would ever seem adequate to them.

In a similar vein, Scott (1998, 4.3) found that subjects in her study of survivors of ritual abuse wanted her to write 'a comprehensive report that validates survivors' realities'. She writes:

> Knowing that the media, legal and psychiatric systems largely characterized them as unreliable witnesses to their own experiences, they were hopeful that 'being researched' could transform their experiential knowledge into legitimate academic discourse. (Scott, 1998, 4.4)

Like Scott, I struggled with the sense that I was failing my subjects. Nothing would maintain my integrity as a sociologist and a person committed to resisting oppression while adequately telling the stories of the persons with whom I had spoken.

Research as Violence to the Self

The intensity with which I connected to my subjects contributed to a feeling they narrated my day–to-day life. I heard their voices in my head—felt the swelling of their emotions within me. I wondered if I could critically reflect on their experiences. To talk about the process of storytelling, to dissect the way in which we construct and reconstruct our histories for anyone, can make subjects feel their stories are somehow less authentic, less 'the truth'. That has never been my intention. As a social constructionist, though, that was the task before me: talk about the nuts and bolts of their meaning making.

Fine (2002) writes that social movements are 'bundles of narratives'. It is not surprising that ex-gay tales—one part recovery narrative, one part salvational testimony—hijacked my mind. I internalized the narratives in ex-gay books, stories, conference sessions, websites and interview transcripts. Three years in, I was able to put into words what I was feeling. Drawing from Rosenberg's (1998) understanding of violence as something that we often do to ourselves, I see this work as violence I inflicted upon myself. A successfully defended dissertation proposal is a bed one makes then lies in. I let myself drown, with perverse abandon, in my work.

I have begun to understand, in the words of Lyn Yates (1997, p. 496), 'the research project as lived dilemma rather than simply as the neat achievement presented in the published report'. Qualitative research requires a great deal of 'emotional labor' (Hochschild, 1983; see also Hubbard et al., 2001). Every aspect of my research incorporated emotional labor: recruiting, interviewing, participant observation, analyzing and writing.

Conclusion

Only with the distance and the fading voices have I been able to engage the data with a critical eye. I remain troubled by this fact, wondering if that distance is a way to dehumanize and break the ties of my earlier connections. My guilt, though, is tempered by the relief I feel from detachment.

When people ask what is coming up on my research agenda, I often say 'something happy'. This research has taken its toll on me, and I am ready for a body of research from which I am emotionally distant. I wonder, though, if the quality of this next work will be anything near my work with the ex-gay movement.

Acknowledgements

The author wishes to thank Avery Brooks Thompkins, Allison Huber, Jennifer Dunn, Robert Benford, Kevin Gillan and anonymous reviewers for their insightful feedback at various stages of the writing of this essay.

References

Burr, G. (1996) Unfinished business: Interviewing family members of critically ill patients, *Nursing Inquiry*, 3(3), pp. 172–177.

Fine, Gary Alan (2002) The storied group: Social movements as 'bundles of narratives', in: J. Davis (Ed.) *Stories Of Change: Narrative and Social Movements*, pp. 229–245 (Albany, NY: State University of New York Press).

Hubbard, Gill, Backett-Milburn, Kathryn & Kemmer, Debbie (2001) Working with emotion: Issues for the researcher in fieldwork and teamwork, *International Journal of Social Research Methodology*, 4(2), pp. 119–137.

Hochschild, Arlie R. (1983) The managed heart: Commercialization of human feeling. Berkeley, CA: University of California Press.

Pickett, Mary, Brennan, Anne Marie Walsh, Greenberg, Helaine S., Licht, Lois & Worrell, Judith Deignan (1994) Use of debriefing techniques to prevent compassion fatigue in research teams, *Nursing Research*, 43(4), pp. 250–252.

Ponticelli, Christy M. (2005) Shades of grey or back to nature?, in: S. Thumma & E. R. Gray (Eds) *Gay Religion*, pp. 153–162 (Lanham, MD: Altamira Press).

Rosenberg, Marshall B. (1998) *Nonviolent Communication* (Encinitas, CA: Puddledancer Press).

Sampson, H., Bloor, M. & Fincham, B. (2008) A price worth paying?, *Sociology*, 42(5), pp. 919–933.

Scott, S. (1998) Here be dragons: Researching the unbelievable, hearing the unthinkable: A feminist sociologist in uncharted territory, *Sociology Research Online*, Available at http://www.socresonline.org.uk/3/3/1.html (accessed 8 January 2011).

Yates, L. (1997) Gender equity and the boys debate: What sort of challenge is it?, *British Journal of Sociology of Education*, 18(3), pp. 337–347.

Index

Note: Page numbers followed by 'n' refer to notes

INDEX